Practical Analysis of COMPOSITE LAMINATES

J.N. Reddy
Department of Mechanical Engineering
Texas A&M University
College Station, Texas

A. Miravete
Department of Mechanical Engineering
University of Zaragoza
Zaragoza, Spain

CRC Press
Boca Raton New York London Tokyo

Library of Congress Cataloging-in-Publication Data

Reddy, J. N. (Junuthula Narasimha), 1945-
Practical analysis of composite laminates / J.N. Reddy and A. Miravete.
p. cm.
Includes bibliographical references and index.
ISBN 0-8493-9401-5 (alk. paper)
1. Composite construction. 2. Laminated materials. 3. Structural analysis (Engineering) I. Miravete, A. (Antonio) II. Title.
TA664.R45 1995
624.1'8--dc20

95-34343
CIP

Direct all inquiries to CRC Press, Inc., 2000 Corporate Blvd., N.W., Boca Raton, Florida 33431.

International Standard Book Number 0-8493-9401-5
Library of Congress Card Number 95-34343
Printed in the United States of America 1 2 3 4 5 6 7 8 9 0
Printed on acid-free paper

PREFACE

Composite materials are increasingly used in aerospace, underwater, and automotive structures. The application of composite materials to engineering components has spurred a major effort to analyze structural components made from them. Composite materials provide unique advantages over their metallic counterparts, but they also present complex and challenging problems to analysts and designers.

The global deformation of laminated composite structures is often characterized by complex coupling between the extension, bending, and shearing modes. Further, due to their low transverse shear stiffness, composite laminates often exhibit significant transverse shear deformation at lower thickness–to–span ratios than do similar homogeneous isotropic plates and shells. To take advantage of the full potential of composite materials, structural analysts and designers must have accurate analysis tools and design methods at their disposal.

The primary aim of this book is to present a summary of the equations governing composite laminates and show a way to analyze most common types of composite structural elements from a practical point of view. A number of experimental results for several types of structures are included, and theoretical and experimental correlations are discussed. A PC based software, *DAC* (Designing Advanced Composites), is discussed in the last chapter. The book serves as a reference for practicing engineers and designers of composite materials. It can also be used as a text book at the graduate level.

The first author expresses his sincere thanks to Dr. Ahmed Khdeir for his research collaboration and contribution to the Lévy type analytical solutions of rectangular laminates (see Chapter 5) and to Mr. G. N. Praveen for the literature review presented in Chapter 6. The second author expresses his thanks to Drs. Stephan W. Tsai and Ran Y. Kim for their discussions on variable thickness structures (see Chapter 7). The second author also wishes to acknowledge research collaboration in composite structures with his colleagues Guido Van den Broeck, Pedro M. Bravo, Emilio Larrode, Miguel A. Jimenez, Juan J. Alba, and Maria J. Casamayor. It is with great pleasure and appreciation that the authors acknowledge the patience in typing of portions of this manuscript by Mrs. Diane Heintzen and Miss Teresa Laborde.

J. N. Reddy
College Station, Texas

A. Miravete
Zaragoza, Spain

ABOUT THE AUTHORS

Professor J. N. Reddy is the inaugural holder of the *Oscar S. Wyatt Endowed Chair* in Mechanical Engineering at Texas A&M University, College Station, Texas. He earned a Ph.D. in Engineering Mechanics, and worked as a Postdoctoral Fellow at the University of Texas at Austin, Research Scientist for Lockheed Missiles and Space Company during 1974–75, and taught at the University of Oklahoma from 1975 to 1980 and Virginia Polytechnic Institute & State University from 1980 to 1992. Dr. Reddy authored numerous papers on the theory and finite element analysis of problems in continuum mechanics, solid and structural mechanics, laminated composite plates and shells, computational fluid mechanics, numerical heat transfer, and applied mathematics. Dr. Reddy is the author of eight other books, including: *An Introduction to the Finite Element Method*, McGraw–Hill, 1984 (second edition, 1993) and *Energy and Variational Methods in Applied Mechanics*, John Wiley & Sons, 1984. Dr. Reddy is the recipient of the *Walter L. Huber Civil Engineering Research Prize* of the American Society of Civil Engineers (ASCE) and the *Worcester Reed Warner Medal* of the American Society of Mechanical Engineers (ASME). Dr. Reddy is a fellow of the American Academy of Mechanics (AAM), the American Society of Civil Engineers (ASCE), the American Society of Mechanical Engineers (ASME), and the Aeronautical Society of India (ASI). Dr. Reddy serves on the editorial boards of numerous journals, including *Journal of Applied Mechanics,* and *International Journal for Numerical Methods in Engineering.* He is the Editor–in–Chief of *Mechanics of Composite Materials and Structures.*

Professor Antonio Miravete is the Chairman of the Department of Mechanical Engineering, University of Zaragoza, Zaragoza, Spain. He earned a Ph.D. in Engineering Mechanics and worked as a Postdoctoral Fellow at the University of Zaragoza. He was invited as a Visiting Scientist in the Non–Metallic Materials Laboratory at Wright Patterson Air Force Base, Dayton, Ohio (1988–89) and in the Aeronautics and Astronautics Department at Stanford University (1992). Dr. Miravete's research interests are primarily in the analysis and design of composite structures. He has several contracts with major industrial companies in Europe (BMW, ASA, GEC Alsthom, Daimler Benz, Deutsche Aerospace, Hexcel, Intermarine, Parabeam, Vetrotex, and others). Dr. Miravete authored numerous papers on analysis and design of composite structures, and theory–experiment correlations. He wrote some books in the field of composite materials and translated the book *Composites Design* written by Stephen W. Tsai into Spanish. He was the Chairman of the *Ninth International Conference on Composite Materials (ICCM/9)*, Madrid, 12–16 July 1993, and edited six volumes of the ICCM/9 proceedings. Dr. Miravete serves on the editorial boards of numerous national and international journals, including *International Journal of Composite Structures, International Journal of Composite Materials and Structures,* and *International Journal of Applied Composite Materials.*

To our wives

Aruna and **Pilar**

CONTENTS

Practical Analysis of COMPOSITE LAMINATES

Chapter One

INTRODUCTION AND PRELIMINARIES

1.1 INTRODUCTION

Composite materials are those that consist of two or more constituent materials that together produce desirable properties for a given application. The traditional mixture of sand and cementing materials and reinforced steel bars embedded in concrete provide familiar examples of composite materials. Fiber–reinforced composite materials are the most commonly used modern composite materials that consist of high strength and high modulus *fibers* in a *matrix* material. In these composites, fibers are the principal load–carrying members, and the matrix material keeps the fibers together, acts as a load–transfer medium between fibers, and protects fibers from being exposed to environment (*e.g.*, moisture, humidity, etc.). It is known that fibers are stiffer and stronger than the same material in bulk form, whereas matrix materials have their usual bulk–form properties. Geometrically, fibers have near crystal–sized diameter and a very high length–to–diameter ratio. Short fibers, called *whiskers*, paradoxically exhibit better properties than long fibers. Long fibers are used in straight form or woven form.

The fibers and matrix materials used in composites are either metallic or nonmetallic. The fiber materials in use are common metals like aluminum, copper, iron, nickel, steel, and titanium, and organic materials like glass, carbon, boron, and graphite materials. New material systems are under continuous development. Fiber–reinforced composite materials for structural applications are often made in the form of a thin layer, called *lamina*. Structural elements are then formed by stacking the layers to desired thickness and properties. Fiber orientation in each lamina and stacking sequence of the layers can be chosen to achieve desired strength and stiffness for a specific application.

1.2 PRESENT STUDY

A study of composite materials for structural applications involves many aspects: the development of material systems, characterization of the material systems, analysis and design of structural elements, and manufacturing of structural components, to name a few. The present study is focussed on the analysis and design aspects of some commonly used structural components, like bars, beams, and plates. The study requires anisotropic elasticity equations, structural theories (*i.e.*, kinematics of deformation), analytical or computational methods to solve the governing equations, and damage/failure theories to predict modes of failures and failure loads. Case studies of design of some commonly used practical structural elements are also included. Emphasis is placed on engineering aspects and simple analysis concepts that are important in the design of practical composite structures.

Following this introduction, some useful mathematical preliminaries are reviewed in this chapter. A review of the basic equations of anisotropic elasticity is presented in Chapter 2. A complete discussion of the mechanical behavior (*i.e.*, constitutive equations) of a composite lamina is presented in Chapter 3. Chapter 4 is devoted to the development of common structural theories of composite laminates. Analytical and numerical (*i.e.*, finite element) solution procedures for composite laminates are presented in Chapter 5. Numerical results for bending, vibration, and buckling of rectangular plates are also included in Chapter 5. Chapter 6 contains an overview of failure theories. Chapters 7 through 9 are devoted to practical analysis of variable thickness structures, crash, and large structures by substructure techniques, respectively. Chapter 10 contains a description of the computer software termed *DAC*.

1.3 MATHEMATICAL PRELIMINARIES

In the analytical study of any physical problem one is required to formulate the governing equations with respect to a conveniently chosen coordinate system. In the study of structures we encounter quantities that are scalars, vectors, and tensors (or vectors of higher order). For the sake of completeness and ready reference, a discussion of these quantities is presented here.

When the basis vectors of a coordinate system are constant, *i.e.*, with fixed lengths and directions, the coordinate system is called a *Cartesian coordinate system.* When the Cartesian system is orthogonal, it is called *rectangular Cartesian.* When the basis vectors are of unit magnitude and orthogonal they are called *orthonormal.* We denote an orthonormal Cartesian basis by

$$(\hat{\mathbf{e}}_1, \hat{\mathbf{e}}_2, \hat{\mathbf{e}}_3) \quad \text{or} \quad (\hat{\mathbf{e}}_x, \hat{\mathbf{e}}_y, \hat{\mathbf{e}}_z)$$

The Cartesian coordinates are denoted by

$$(x_1, x_2, x_3) \quad \text{or} \quad (x, y, z)$$

The familiar rectangular Cartesian coordinate system is shown in Fig. 1.3–1. We shall always use right–handed coordinate systems.

It is convenient to abbreviate a summation of terms by understanding that repeated index means summation over all values of that index. For example the component form of vector $\mathbf{A}$

$$\mathbf{A} = a^1\hat{\mathbf{e}}_1 + a^2\hat{\mathbf{e}}_2 + a^3\hat{\mathbf{e}}_3 \qquad (1.3-1)$$

where $(\hat{\mathbf{e}}_1, \hat{\mathbf{e}}_2, \hat{\mathbf{e}}_3)$ are basis vectors and (a^1, a^2, a^3) are the components of $\mathbf{A}$, can be expressed as

$$\mathbf{A} = \sum_{j=1}^{3} a^j\hat{\mathbf{e}}_j = a^j\hat{\mathbf{e}}_j \qquad (1.3-2)$$

where the summation symbol is omitted with the understanding that the expression is summed over the range of the repeated index: $j = 1, 2, 3$. The repeated index is a *dummy index* in the sense that any other symbol that is not already used in that expression can be employed:

$$\mathbf{A} = a^j\hat{\mathbf{e}}_j = a^k\hat{\mathbf{e}}_k = a^m\hat{\mathbf{e}}_m$$

The range of summation is known in the context of the discussion. A *free index* is an index that is not repeated.

The *Kronecker delta*, denoted δ_{ij}, is defined as

$$\delta_{ij} = \begin{cases} 0, & i \neq j \\ 1, & i = j \end{cases} \qquad (1.3-3)$$

for any fixed values of i and j. The Kronecker delta symbol can be used to advantage in expressing vector equations in component form. For example, the nine orthogonality conditions among the unit basis vectors $(\hat{\mathbf{e}}_1, \hat{\mathbf{e}}_2, \hat{\mathbf{e}}_3)$ of a rectangular Cartesian system are

$$\begin{aligned} &\hat{\mathbf{e}}_1 \cdot \hat{\mathbf{e}}_1 = 1, \quad \hat{\mathbf{e}}_1 \cdot \hat{\mathbf{e}}_2 = 0, \quad \hat{\mathbf{e}}_1 \cdot \hat{\mathbf{e}}_3 = 0 \\ &\hat{\mathbf{e}}_2 \cdot \hat{\mathbf{e}}_1 = 0, \quad \hat{\mathbf{e}}_2 \cdot \hat{\mathbf{e}}_2 = 1, \quad \hat{\mathbf{e}}_2 \cdot \hat{\mathbf{e}}_3 = 0 \\ &\hat{\mathbf{e}}_3 \cdot \hat{\mathbf{e}}_1 = 0, \quad \hat{\mathbf{e}}_3 \cdot \hat{\mathbf{e}}_2 = 0, \quad \hat{\mathbf{e}}_3 \cdot \hat{\mathbf{e}}_3 = 1 \end{aligned} \qquad (1.3-4a)$$

These relations can be expressed as

$$\hat{\mathbf{e}}_i \cdot \hat{\mathbf{e}}_j = \delta_{ij} \qquad (1.3-4b)$$

Note that both i and j in Eq. (4) are free indices, *i.e.*, they cannot be renamed on one side without renaming on the other side by the same indices:

$$\hat{\mathbf{e}}_k \cdot \hat{\mathbf{e}}_\ell = \delta_{k\ell}, \quad \hat{\mathbf{e}}_m \cdot \hat{\mathbf{e}}_n = \delta_{mn}$$

Next consider the nine cross-product (or vector product) relations among the basis vectors $(\hat{\mathbf{e}}_1, \hat{\mathbf{e}}_2, \hat{\mathbf{e}}_3)$:

$$\begin{aligned} &\hat{\mathbf{e}}_1 \times \hat{\mathbf{e}}_1 = \mathbf{0}, \quad \hat{\mathbf{e}}_1 \times \hat{\mathbf{e}}_2 = \hat{\mathbf{e}}_3, \quad \hat{\mathbf{e}}_1 \times \hat{\mathbf{e}}_3 = -\hat{\mathbf{e}}_2 \\ &\hat{\mathbf{e}}_2 \times \hat{\mathbf{e}}_1 = -\hat{\mathbf{e}}_3, \quad \hat{\mathbf{e}}_2 \times \hat{\mathbf{e}}_2 = \mathbf{0}, \quad \hat{\mathbf{e}}_2 \times \hat{\mathbf{e}}_3 = \hat{\mathbf{e}}_1 \\ &\hat{\mathbf{e}}_3 \times \hat{\mathbf{e}}_1 = \hat{\mathbf{e}}_2, \quad \hat{\mathbf{e}}_3 \times \hat{\mathbf{e}}_2 = -\hat{\mathbf{e}}_1, \quad \hat{\mathbf{e}}_3 \times \hat{\mathbf{e}}_3 = \mathbf{0} \end{aligned} \qquad (1.3-5a)$$

These relations can be written in compact form as

$$\hat{\mathbf{e}}_i \times \hat{\mathbf{e}}_j = \epsilon_{ijk}\hat{\mathbf{e}}_k \qquad (1.3-5b)$$

where ϵ_{ijk} is the *permutation symbol*, defined as $(i \neq j \neq k)$

$$\epsilon_{ijk} = \begin{cases} 1, & \text{if } i,\ j,\ k \text{ are in cyclic order and not repeated} \\ 0, & \text{if any } i,\ j,\ k \text{ are repeated} \\ -1, & \text{if } i,\ j,\ k \text{ are not in cyclic order and not repeated} \end{cases} \qquad (1.3-6)$$

In Eq. (5) i and j are free indices, and k is a dummy index, and summation convention is used.

The symbols δ_{ij} and ϵ_{ijk} are related by the $\epsilon - \delta$ *identity*:

$$\epsilon_{ijk}\epsilon_{mnk} = \delta_{im}\delta_{jn} - \delta_{in}\delta_{jm} \qquad (1.3-7)$$

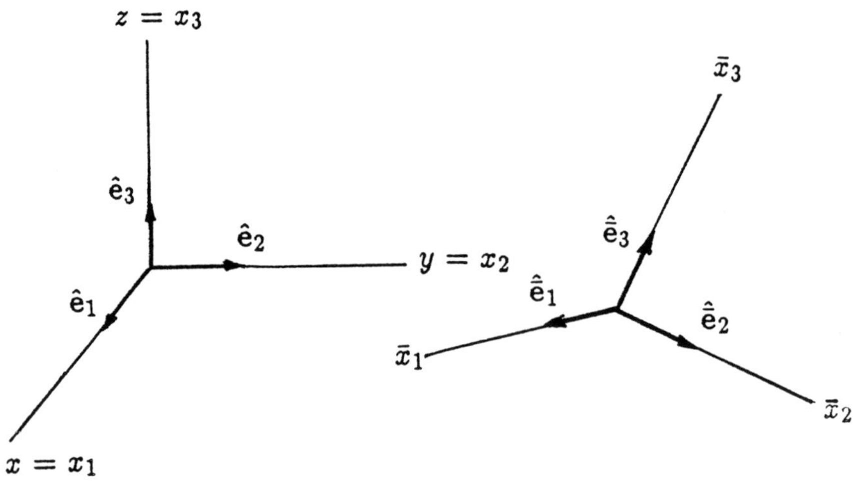

Figure 1.3–1: Rectangular Cartesian coordinate systems.

Quantities that require one or more basis vectors to represent them are generally called *tensors*. A vector is a first–order tensor. A *dyad* is an entity of two vectors standing next to each other with no operation between them, and it is a second–order tensor. Examples of a second–order tensor are provided by stress and strain at a point. Stress at a point, for example, depends on the plane on which it is calculated and the direction in which it is acting. Thus, stress is a dyad, requiring specification of two directions and a magnitude.

A dyad is said to be *symmetric* if it is equal to its transpose:

$$\mathbf{\Phi} = \mathbf{\Phi}^T \tag{1.3 – 8}$$

where $(\cdot)^T$ denotes the transpose of the enclosed quantity. Equation (8) implies that the nine components of a symmetric dyad are related:

$$\begin{aligned} \varphi_{ij}\hat{\mathbf{e}}_i\hat{\mathbf{e}}_j &= \varphi_{ij}\hat{\mathbf{e}}_j\hat{\mathbf{e}}_i \\ &= \varphi_{ji}\hat{\mathbf{e}}_i\hat{\mathbf{e}}_j \end{aligned}$$

or

$$\varphi_{ij} = \varphi_{ji} \quad \text{(symmetric)} \tag{1.3 – 9}$$

Thus only six components of a symmetric dyad are independent.

A dyad is *antisymmetric* if it is equal to the negative of its transpose:

$$\mathbf{\Phi} = -\mathbf{\Phi}^T \tag{1.3 – 10}$$

The nine components of an antisymmetric dyad are related as follows:

$$\begin{aligned} \varphi_{ij}\hat{\mathbf{e}}_i\hat{\mathbf{e}}_j &= -\varphi_{ij}\hat{\mathbf{e}}_j\hat{\mathbf{e}}_i \\ &= -\varphi_{ji}\hat{\mathbf{e}}_i\hat{\mathbf{e}}_j \end{aligned}$$

or

$$\varphi_{ij} = -\varphi_{ji} \quad \text{(antisymmetric)} \tag{1.3 – 11a}$$

Clearly, when $i = j$, we must have

$$\varphi_{11} = -\varphi_{11} = 0$$

$$\varphi_{22} = -\varphi_{22} = 0$$

$$\varphi_{33} = -\varphi_{33} = 0 \qquad (1.3-11b)$$

Hence an antisymmetric dyad has only *three* independent components.

In the analysis of laminated structures, one is required to refer all quantities – strains, stresses, and material stiffnesses – to a common coordinate system. To establish the rules of the transformation of the components of vectors (*e.g.*, displacements) and tensors (*e.g.*, strains and stresses), we consider two coordinate systems, $(\bar{x}_1, \bar{x}_2, \bar{x}_3)$ and (x_1, x_2, x_3), that are related by the following transformation

$$\begin{aligned} x_1 &= x_1(\bar{x}_1, \bar{x}_2, \bar{x}_3) \\ x_2 &= x_2(\bar{x}_1, \bar{x}_2, \bar{x}_3) \\ x_3 &= x_3(\bar{x}_1, \bar{x}_2, \bar{x}_3) \end{aligned} \qquad (1.3-12)$$

The inverse transformation is

$$\begin{aligned} \bar{x}_1 &= \bar{x}_1(x_1, x_2, x_3) \\ \bar{x}_2 &= \bar{x}_2(x_1, x_2, x_3) \\ \bar{x}_3 &= \bar{x}_3(x_1, x_2, x_3) \end{aligned} \qquad (1.3-13)$$

The bases in the two systems are denoted $(\hat{\mathbf{e}}_1, \hat{\mathbf{e}}_2, \hat{\mathbf{e}}_3)$ and $(\hat{\bar{\mathbf{e}}}_1, \hat{\bar{\mathbf{e}}}_2, \hat{\bar{\mathbf{e}}}_3)$ (see Figure 1.3–1).

The relations between the bases of the barred (new) and unbarred (old) coordinate systems can be written as

$$\hat{\bar{\mathbf{e}}}_i = (\hat{\bar{\mathbf{e}}}_i \cdot \hat{\mathbf{e}}_k)\, \hat{\mathbf{e}}_k = a_{ik}\hat{\mathbf{e}}_k \qquad (1.3-14)$$

We also have the following inverse relation

$$\hat{\mathbf{e}}_j = (\hat{\mathbf{e}}_j \cdot \hat{\bar{\mathbf{e}}}_k)\, \hat{\bar{\mathbf{e}}}_k = a_{kj}\hat{\bar{\mathbf{e}}}_k \qquad (1.3-15)$$

From Eqs. (14) and (15), we note that

$$a_{ik}a_{jk} = \delta_{ij} \quad \text{or} \quad [A][A]^T = [I] \qquad (1.3-16)$$

which implies that $[A]^T$ is the inverse of $[A]$.

The transformations (14) and (15) between two orthogonal sets of bases also hold for the respective coordinates:

$$\bar{x}_i = a_{ik}x_k \quad \text{or} \quad \{\bar{x}\} = [A]\{x\} \qquad (1.3-17a)$$

$$x_j = a_{kj}\bar{x}_k \quad \text{or} \quad \{x\} = [A]^T\{\bar{x}\} \qquad (1.3-17b)$$

For example, suppose that $(\bar{x}_1, \bar{x}_2, \bar{x}_3)$ is obtained from (x_1, x_2, x_3) by rotating the x_1x_2–plane counterclockwise by an angle θ about the x_3–axis. Then the two sets of coordinates are related by [cf. Eq. (13)]

$$\begin{Bmatrix} \bar{x}_1 \\ \bar{x}_2 \\ \bar{x}_3 \end{Bmatrix} = \begin{bmatrix} \cos\theta & \sin\theta & 0 \\ -\sin\theta & \cos\theta & 0 \\ 0 & 0 & 1 \end{bmatrix} \begin{Bmatrix} x_1 \\ x_2 \\ x_3 \end{Bmatrix} \qquad (1.3-18)$$

Next we establish transformation relations among the components of a second–order tensor in two different coordinate systems. A second–order tensor $\mathbf{\Phi}$ can be expressed in two different coordinate systems using the corresponding bases. With respect to the basis $(\hat{\bar{\mathbf{e}}}_1, \hat{\bar{\mathbf{e}}}_2, \hat{\bar{\mathbf{e}}}_3)$, we have

$$\mathbf{\Phi} = \bar{\varphi}_{mn}\hat{\bar{\mathbf{e}}}_m\hat{\bar{\mathbf{e}}}_n \qquad (1.3-19)$$

and with respect to the basis $(\hat{\mathbf{e}}_1, \hat{\mathbf{e}}_2, \hat{\mathbf{e}}_3)$ it is expressed as

$$\mathbf{\Phi} = \varphi_{ij}\hat{\mathbf{e}}_i\hat{\mathbf{e}}_j \qquad (1.3-20)$$

Using Eq. (14) for $\hat{\mathbf{e}}_i$ and $\hat{\mathbf{e}}_j$ in Eq. (20) we arrive at

$$\mathbf{\Phi} = \varphi_{ij}\ a_{mi}a_{nj}\ \hat{\bar{\mathbf{e}}}_m\hat{\bar{\mathbf{e}}}_n \tag{1.3 – 21}$$

Comparing Eq. (21) with Eq. (19), we arrive at the relation

$$\bar{\varphi}_{mn} = a_{mi}a_{nj}\ \varphi_{ij} \tag{1.3 – 22a}$$

or in matrix form, we have

$$[\bar{\varphi}] = [A][\varphi][A]^T \tag{1.3 – 22b}$$

The inverse relation can be derived using the property $[A]^{-1} = [A]^T$. Premultiplying both sides of Eq. (22b) with $[A]^{-1} = [A]^T$ and postmultiplying both sides of the resulting equation with $([A]^T)^{-1} = [A]$, we obtain the result

$$[\varphi] = [A]^T[\bar{\varphi}][A] \tag{1.3 – 23}$$

The transformation law (22a) is often taken to be the *definition* of a second–order tensor. In other words, $\mathbf{\Phi}$ is a second–order tensor if and only if its components transform according to Eq. (22a). In general, an n–th order tensor transforms according to the formula

$$\bar{\varphi}_{mnpq\ldots} = \varphi_{ijk\ell\ldots}a_{mi}a_{nj}a_{pk}a_{q\ell}\cdots \tag{1.3 – 24}$$

EXERCISE PROBLEMS

1.1 Verify the following identities by actual expansion (*i.e.*, sum repeated indices):

(a) $\delta_{ii} = 3$

(b) $\epsilon_{ijk}\epsilon_{ijk} = 6$

(c) $A_iA_j\epsilon_{ijk} = 0$

1.2 Use Eqs. (1.3–14) and (1.3–15) to establish the identity in Eq. (1.3–16).

1.3 Show that φ_{ij} are symmetric if

$$\varphi_{ij}\epsilon_{ijk} = 0 \quad \text{for all} \quad k$$

1.4 Compute the transformed coefficients of a second–order tensor $\boldsymbol{\Phi}$ under the transformation defined by Eq. (1.3–18).

REFERENCES FOR ADDITIONAL READING

1. Bowen, R. M. and Wang, C. C., *Introduction to Vectors and Tensors*, Vols. I and II, Plenum Press, New York, 1976.

2. Jeffreys, H., *Cartesian Tensors*, Cambridge University Press, London, 1965.

3. Reddy, J. N. and Rasmussen, M. L., *Advanced Engineering Analysis*, John Wiley & Sons, New York, 1982; reprinted by Krieger, Melbourne, FL, 1990.

4. Reddy, J. N., *Mechanics of Composite Materials: Theory and Analysis,* John Wiley & Sons, New York, 1996, in press.

Chapter Two

REVIEW OF GOVERNING EQUATIONS

2.1 CLASSIFICATION OF EQUATIONS

The objective of this chapter is to review the governing equations of elasticity. In much of this study we shall be concerned with the motion or equilibrium of solid bodies that are made of composite materials, under the action of externally applied mechanical and/or thermal loads.

The equations governing the motion of a solid body can be classified into four basic categories:

(1) Kinematics (strain–displacement equations)

(2) Kinetics (conservation of linear and angular momenta)

(3) Thermodynamic principles (first and second laws of thermodynamics)

(4) Constitutive equations (stress–strain relations)

Kinematics is study of the geometry of motion and deformation of a body without consideration of the forces causing the motion. Kinetics is the study of the static or dynamic equilibrium of forces acting on a body. The thermodynamic principles are concerned with the relations among heat, work, and thermodynamic properties of the body. The constitutive equations describe the constitutive behavior of the body and relate the dependent variables introduced in the kinetic description to those in the kinematic and thermodynamic descriptions. These equations are supplemented by appropriate boundary and initial conditions of the problem.

In the following sections we will review the strain–displacement relations, equations of motion, and constitutive relations of a linear elastic body. Thermodynamic principles will not be reviewed here, and they are discussed when needed. Since the objective here is to review rather than to develop the basic equations of mechanics, only a brief discussion of the derivations will be given. For a more detailed account of the theory of a continuous medium, the reader is referred to a number of books suggested in the bibliography at the end of the Chapter.

2.2 KINEMATICS

In solid mechanics the reference configuration C° is often chosen as the unstressed state of the solid body, Ω. The coordinates $\mathbf{X} = (X_1, X_2, X_3)$ are called the *material coordinates* of the particle X. In the so–called Lagrangian description, commonly used in solid mechanics, the coordinate system $\mathbf{X}$ is fixed on a given body of matter in its undeformed configuration, and its position $\mathbf{x}$ any time is referred to the material coordinate X_i :

$$\mathbf{x} = \mathbf{x}(X_1, X_2, X_3, t) \tag{2.2 – 1}$$

The motion described by Eq. (1) gives the position $\mathbf{x}$ occupied at time t by a particle X whose position in the reference configuration was $\mathbf{X}$. Thus, during a motion of a body B, a representative particle X occupies a succession of points that together form a curve in euclidean space. This curve is called the path of X and is given parametrically by Eq. (1), with

$$\mathbf{x} = \mathbf{x}(t), \quad \mathbf{x}(0) = \mathbf{X} \tag{2.2 – 2}$$

The word *deformation* refers to displacements and changes in the geometry experienced by particles in a body, Ω. Referred to a rectangular Cartesian frame of reference (X_1, X_2, X_3), every particle X in the body has a set of coordinates $\mathbf{X}$. When the body is deformed under the action of external forces, particle X, which occupies place $\mathbf{X}$ in the undeformed body, moves to a new position $\mathbf{x} = (x_1, x_2, x_3)$. The displacement of the particle X is given by

$$\mathbf{u} = \mathbf{x} - \mathbf{X} \quad \text{or} \quad u_i = x_i - X_i \tag{2.2-3}$$

If the displacement of every particle in the body is known, we can construct the deformed configuration C from the reference (or undeformed) configuration C°.

If the displacements are expressed in terms of the coordinates X_i in the reference configuration (the Lagrangian description), we have

$$u_i(X_1, X_2, X_3, t) = x_i(X_1, X_2, X_3, t) - X_i \tag{2.2-4}$$

The strain tensor $\mathbf{E}$, known as the *Green–Lagrange strain tensor*, is defined by

$$\mathbf{E} = \frac{1}{2}\left[\nabla\mathbf{u} + (\nabla\mathbf{u})^T + (\nabla\mathbf{u}) \cdot (\nabla\mathbf{u})^T\right] \tag{2.2-5}$$

where $(\cdot)^T$ denotes transpose of the enclosed quantity, and ∇ is the gradient operator with respect to the coordinate $\mathbf{X}$. Note that, by definition, the strain tensor is symmetric. Hence, there are only six independent components of strain. The Cartesian components of the Green–Lagrange strain tensor are

$$E_{jk} = \frac{1}{2}\left(\frac{\partial u_j}{\partial X_k} + \frac{\partial u_k}{\partial X_j} + \frac{\partial u_m}{\partial X_j}\frac{\partial u_m}{\partial X_k}\right) \tag{2.2-6a}$$

or in explicit form ($E_{21} = E_{12}, E_{31} = E_{13}, E_{32} = E_{23}$),

$$E_{11} = \frac{\partial u_1}{\partial X_1} + \frac{1}{2}\left[\left(\frac{\partial u_1}{\partial X_1}\right)^2 + \left(\frac{\partial u_2}{\partial X_1}\right)^2 + \left(\frac{\partial u_3}{\partial X_1}\right)^2\right]$$

$$E_{22} = \frac{\partial u_2}{\partial X_2} + \frac{1}{2}\left[\left(\frac{\partial u_1}{\partial X_2}\right)^2 + \left(\frac{\partial u_2}{\partial X_2}\right)^2 + \left(\frac{\partial u_3}{\partial X_2}\right)^2\right]$$

$$E_{33} = \frac{\partial u_3}{\partial X_3} + \frac{1}{2}\left[\left(\frac{\partial u_1}{\partial X_3}\right)^2 + \left(\frac{\partial u_2}{\partial X_3}\right)^2 + \left(\frac{\partial u_3}{\partial X_3}\right)^2\right]$$

$$
\begin{aligned}
E_{12} &= \frac{1}{2}\left(\frac{\partial u_1}{\partial X_2} + \frac{\partial u_2}{\partial X_1} + \frac{\partial u_1}{\partial X_1}\frac{\partial u_1}{\partial X_2} + \frac{\partial u_2}{\partial X_1}\frac{\partial u_2}{\partial X_2} + \frac{\partial u_3}{\partial X_1}\frac{\partial u_3}{\partial X_2}\right)\\
E_{13} &= \frac{1}{2}\left(\frac{\partial u_1}{\partial X_3} + \frac{\partial u_3}{\partial X_1} + \frac{\partial u_1}{\partial X_1}\frac{\partial u_1}{\partial X_3} + \frac{\partial u_2}{\partial X_1}\frac{\partial u_2}{\partial X_3} + \frac{\partial u_3}{\partial X_1}\frac{\partial u_3}{\partial X_3}\right)\\
E_{23} &= \frac{1}{2}\left(\frac{\partial u_2}{\partial X_3} + \frac{\partial u_3}{\partial X_2} + \frac{\partial u_1}{\partial X_2}\frac{\partial u_1}{\partial X_3} + \frac{\partial u_2}{\partial X_2}\frac{\partial u_2}{\partial X_3} + \frac{\partial u_3}{\partial X_2}\frac{\partial u_3}{\partial X_3}\right)
\end{aligned} \tag{2.2 - 6b}
$$

The strain components defined in Eq. (5) are called finite strain components because no assumption concerning the smallness (compared to unity) of the strains is made. If the displacement gradients are so small $|\nabla \mathbf{u}| << 1$ or $|u_{i,j}| << 1$ that their squares and products are negligible and the difference between $\frac{\partial u_i}{\partial x_i}$ and $\frac{\partial u_i}{\partial X_i}$ vanishes, then the Green–Lagrange strain tensor reduces to the Almansi (or Euler's) infinitesimal strain tensor:

$$
\varepsilon = \frac{1}{2}\left[\nabla_{\mathbf{x}}\mathbf{u} + (\nabla_{\mathbf{x}}\mathbf{u})^T\right] \tag{2.2 - 7a}
$$

or, in cartesian component form,

$$
\varepsilon_{ij} = \frac{1}{2}\left(\frac{\partial u_i}{\partial x_j} + \frac{\partial u_j}{\partial x_i}\right) \tag{2.2 - 7b}
$$

The explicit form of the infinitesimal strain components is given by

$$
\begin{aligned}
\varepsilon_{11} &= \frac{\partial u_1}{\partial x_1}\\
\varepsilon_{22} &= \frac{\partial u_2}{\partial x_2}\\
\varepsilon_{33} &= \frac{\partial u_3}{\partial x_3}\\
2\varepsilon_{12} &= \frac{\partial u_1}{\partial x_2} + \frac{\partial u_2}{\partial x_1}\\
2\varepsilon_{13} &= \frac{\partial u_1}{\partial x_3} + \frac{\partial u_3}{\partial x_1}\\
2\varepsilon_{23} &= \frac{\partial u_2}{\partial x_3} + \frac{\partial u_3}{\partial x_2}
\end{aligned} \tag{2.2 - 8}
$$

The engineering shear strains are defined as two times the tensorial shear strain components

$$\gamma_{12} = 2\varepsilon_{12} \ , \ \gamma_{13} = 2\varepsilon_{13} \ , \gamma_{23} = 2\varepsilon_{23} \tag{2.2 - 9}$$

2.3 KINETICS

2.3–1 Stress

Forces acting on a material body can be classified as internal and external. Internal forces are those that resist the tendency of one part of the body to be separated from another part. External forces are those exerted on the body by external means. External forces can be classified into two kinds: body forces and surface forces. *Body forces* act on the elements of volume or mass inside the body. Examples of body forces are gravitational and magnetic forces. *Surface forces* are contact forces acting on the bounding surface of the body. Examples of surface forces are provided by applied forces and contact forces exerted by one body on another.

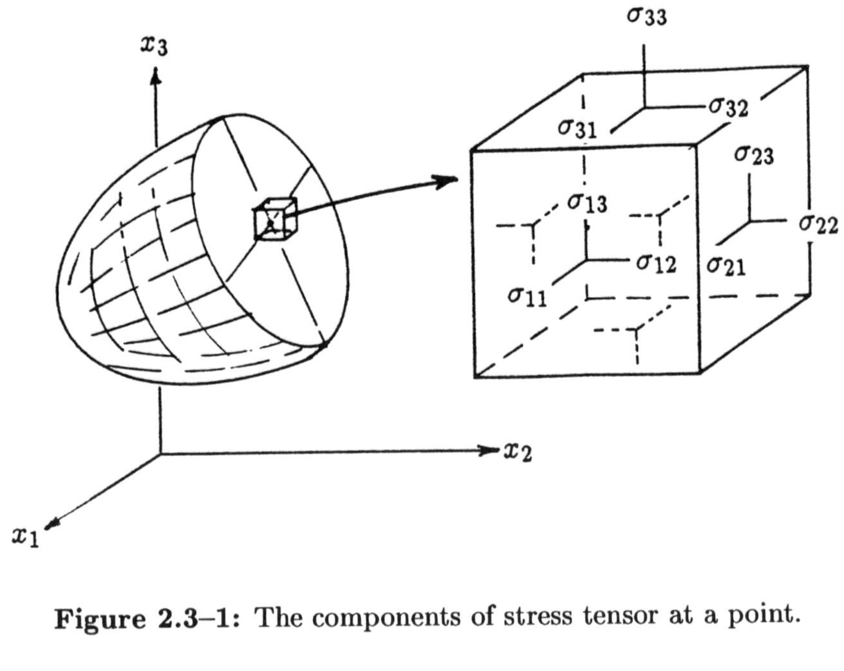

Figure 2.3–1: The components of stress tensor at a point.

Stress at a point is a measure of force per unit area. The force per unit area acting on an elemental area Δs is called the *stress vector* acting on the element. The concept also applies to the surface created by slicing the body with a plane. The stress vector on Δs is the vector $\Delta \mathbf{f}/\Delta s$, which has the same direction as $\Delta \mathbf{f}$ but with magnitude equals to $|\Delta \mathbf{f}|/\Delta s$. The stress vector at a point P on Δs is given by

$$\mathbf{T} = \lim_{\Delta s \to 0} \frac{\Delta \mathbf{f}}{\Delta s} \qquad (2.3-1)$$

The stress vectors on three mutually perpendicular planes, say planes perpendicular to the rectangular coordinate axes, will each have three components — all at the same point. Let $\mathbf{T}^{(i)}$ denote the stress vector at point P on a plane perpendicular to the x_i–axis (see Fig. 2.3–1a). Each vector $\mathbf{T}^{(i)}$ can be resolved into components along the coordinate lines,

$$\mathbf{T}^{(i)} = \sigma_{ij}\hat{\mathbf{e}}_j \ , \ (i \ = \ 1,2,3) \qquad (2.3-2)$$

where σ_{ij} denotes the components of the stress vector $\mathbf{T}^{(i)}$ along the x_j–direction, and $\hat{\mathbf{e}}_j$ is the unit base vector along the x_j–coordinate. Thus σ_{ij} denotes the stress component along the x_j–coordinate and acting on a plane perpendicular to the x_i–axis. The quantities σ_{ij} are the components of a second–order tensor (*i.e.*, dyad), called the Cauchy stress tensor, σ. The stress components σ_{ij} are shown on three perpendicular planes in Fig. 2.3–1b. It should be noted that the cube that is used to indicate the stress components has no dimensions; it is a point cube so that all nine components are acting at the point P.

It can be shown, using the linear momentum principle, that the traction vector $\mathbf{T}^{(\hat{\mathbf{n}})}$ on a plane with unit normal $\hat{\mathbf{n}}$ is related to the stress tensor σ by the relation

$$\mathbf{T}^{(\hat{\mathbf{n}})} = \hat{\mathbf{n}} \cdot \sigma = \sigma^T \cdot \hat{\mathbf{n}} \ \text{ or } \ \{T^{(\hat{\mathbf{n}})}\} = [\sigma]^T\{n\} \qquad (2.3-3)$$

where $[\sigma]^T$ denotes the transpose of $[\sigma]$. Equation (3) is called the *Cauchy stress formula.*

In the above discussion it is understood that stress at a point in a deformed body is measured as the force per unit area in the deformed body. The area element Δs in the deformed body corresponds to an area element ΔS in the reference configuration (in much the same way as $\mathbf{x}$ is the position of a material particle in the deformed body whose position in the reference configuration was $\mathbf{X}$). Here $\mathbf{T}^{(n)}$ is *measured* with respect to the area element in the deformed body. Stress vectors can be *referred* to either the reference coordinate system (X_1, X_2, X_3) or the spatial coordinate system (x_1, x_2, x_3). The stress that is measured in the deformed configuration C and referred to the coordinate system in the deformed body is called the *Cauchy stress* or *true stress*, σ. The components σ_{ij} can be interpreted as the j–th component of the force per unit area in the current configuration C acting on a surface segment whose outward normal at $\mathbf{x}$ is in the i–th direction.

A suitable strain measure to use with the Cauchy stress tensor σ would be the infinitesimal (or the Euler) strain tensor ε. Since we defined strain tensor $\mathbf{E}$ as a function of the material point $\mathbf{X}$ in a reference state, the stress must also be expressed as a function of the material point. The stress measure that is used in nonlinear analysis of solid bodies is the *second Piola–Kirchhoff stress tensor* $\mathbf{S}$ (S_{ij}), which is measured in the deformed body but referred to the material coordinates, X_i. The second Piola–Kirchhoff stress tensor is symmetric whenever the Cauchy stress tensor is symmetric. For small deformation problems, the difference between the two measures of stress disappears. Since the scope of this book is limited to linear analyses, we work with stress tensor σ.

2.3–2 Equations of Motion

Consider a given mass of a material body, instantaneously occupying a volume Ω bounded by a surface Γ. Suppose that the body is acted upon by external forces $\mathbf{T}$ per unit surface area and $\mathbf{f}$ per unit volume. The linear momentum principle states that the rate of change of the total momentum of a given continuous medium equals the vector sum of all the external forces acting on the medium V that initially occupied V_o, provided Newton's Third Law of action and reaction governs the internal forces. The principle leads to the following result:

$$\nabla \cdot \sigma + \mathbf{f} = \rho \frac{\partial^2 \mathbf{u}}{\partial t^2} \qquad (2.3-4)$$

where ρ denotes the material density of the body (measured per unit volume). In rectangular Cartesian component form, Eq. (4) can be written as

$$\frac{\partial \sigma_{ji}}{\partial x_j} + f_i = \rho \frac{\partial^2 u_i}{\partial t^2} \quad (i = 1, 2, 3) \qquad (2.3-5)$$

where summation on repeated indices is implied (*i.e.*, sum on the repeated index j). Expanding Eq. (5), we have

$$\frac{\partial \sigma_{11}}{\partial x_1} + \frac{\partial \sigma_{21}}{\partial x_2} + \frac{\partial \sigma_{31}}{\partial x_3} + f_1 = \rho \frac{\partial^2 u_1}{\partial t^2} \qquad (2.3-6)$$

$$\frac{\partial \sigma_{12}}{\partial x_1} + \frac{\partial \sigma_{22}}{\partial x_2} + \frac{\partial \sigma_{32}}{\partial x_3} + f_2 = \rho \frac{\partial^2 u_2}{\partial t^2} \qquad (2.3-7)$$

$$\frac{\partial \sigma_{13}}{\partial x_1} + \frac{\partial \sigma_{23}}{\partial x_2} + \frac{\partial \sigma_{33}}{\partial x_3} + f_3 = \rho \frac{\partial^2 u_3}{\partial t^2} \qquad (2.3-8)$$

For other coordinate systems, the equations of motion can be derived from Eq. (4) by expressing σ, $\mathbf{f}$, $\mathbf{u}$, and ∇ in that coordinate system (see Problems 2.9 and 2.10).

In the absence of any body (*i.e.*, distributed) moments, the moment of momentum (or conservation of angular momentum) principle leads to the symmetry of the stress tensor

$$\sigma = \sigma^T \quad (\sigma_{ij} = \sigma_{ji}) \qquad (2.3-9)$$

Thus there are only six independent components of the stress tensor.

Note that the equations of motion or equilibrium contain three equations relating six stress components and therefore cannot be solved for all six components. Additional equations are required.

These include the strain–displacement relations discussed in Section 2.2 and constitutive relations or stress–strain relations to be discussed in the next section.

2.4 CONSTITUTIVE EQUATIONS

The kinematic relations (*i.e.*, strain–displacement equations) and the mechanical and thermodynamic principles are applicable to any continuum irrespective of its physical constitution. Here we consider relations among the stresses and strains that characterize the mechanical response of a material. These equations are called *constitutive equations.* The formulation of the constitutive equations for a given material is guided by certain rules. We will not discuss them here, but we will review the linear constitutive relations for solids undergoing small deformations.

A material body is said to be *ideally elastic* when the body recovers, under isothermal conditions, its original form completely upon removal of the forces causing deformation, and there is a one to one relationship between the state of stress and state of strain. Since isothermal conditions are assumed, the constitutive equations do not include creep at constant stress and stress relaxation at constant strain. Thus, the material coefficients that specify the constitutive relationship between the stress and strain components are assumed to be constant during the deformation. This does not automatically imply that we neglect temperature effects on deformation. We must account for the thermal expansion of the material, which can produce strains or stresses as large as those produced by the applied mechanical forces. Here we discuss the constitutive equations for the isothermal case, and the thermal expansion effects will be considered in the sequel.

The generalized Hooke's law relates the nine components of stress to the nine components of strain,

$$\sigma_{ij} = C_{ijk\ell}\varepsilon_{k\ell} \qquad (2.4-1)$$

where $\varepsilon_{k\ell}$ are the infinitesimal strain components, σ_{ij} are the Cauchy stress components, and $C_{ijk\ell}$ are the material coefficients. The nine equations in Eq. (1) contain 81 coefficients. However, due to the symmetry of both σ_{ij} (for nonpolar case) and $\varepsilon_{k\ell}$, it follows that

$$C_{ijk\ell} = C_{jik\ell} \ , \ C_{ijk\ell} = C_{ij\ell k} \qquad (2.4-2)$$

and there are only 36 elastic coefficients or moduli.

Equation (1) can be expressed in the following convenient form using the single–subscript notation for stress and strain components and double–subscript notation for the elastic constants:

$$\begin{Bmatrix} \sigma_1 \\ \sigma_2 \\ \sigma_3 \\ \sigma_4 \\ \sigma_5 \\ \sigma_6 \end{Bmatrix} = \begin{bmatrix} C_{11} & C_{12} & C_{13} & C_{14} & C_{15} & C_{16} \\ C_{21} & C_{22} & C_{23} & C_{24} & C_{25} & C_{26} \\ C_{31} & C_{32} & C_{33} & C_{34} & C_{35} & C_{36} \\ C_{41} & C_{42} & C_{43} & C_{44} & C_{45} & C_{46} \\ C_{51} & C_{52} & C_{53} & C_{54} & C_{55} & C_{56} \\ C_{61} & C_{62} & C_{63} & C_{64} & C_{65} & C_{66} \end{bmatrix} \begin{Bmatrix} \varepsilon_1 \\ \varepsilon_2 \\ \varepsilon_3 \\ \varepsilon_4 \\ \varepsilon_5 \\ \varepsilon_6 \end{Bmatrix} \qquad (2.4-3a)$$

or

$$\sigma_i = C_{ij}\varepsilon_j \qquad (2.4-3b)$$

where C_{ij} are the elastic coefficients, and

$$\sigma_1 = \sigma_{11} \ , \ \sigma_2 = \sigma_{22} \ , \ \sigma_3 = \sigma_{33} \ , \ \sigma_4 = \sigma_{23} \ , \ \sigma_5 = \sigma_{13} \ , \ \sigma_6 = \sigma_{12}$$

$$\varepsilon_1 = \varepsilon_{11} \ , \ \varepsilon_2 = \varepsilon_{22} \ , \ \varepsilon_3 = \varepsilon_{33} \ , \ \varepsilon_4 = 2\varepsilon_{23} \ , \ \varepsilon_5 = 2\varepsilon_{13} \ , \ \varepsilon_6 = 2\varepsilon_{12} \qquad (2.4-4)$$

The single subscript notation (4) used for stresses and strains is called *contracted notation.*

The number of independent material coefficients can be reduced from 36 to 21 if we assume that there exists a potential function W such that

$$\sigma_i = \frac{\partial W}{\partial \epsilon_i} \qquad (2.4-5)$$

The number can be further reduced if the material exhibits certain material symmetries. These cases will be discussed in Chapter 3.

2.5 SUMMARY

The strain–displacement equations and equations of motion in Cartesian component form, under small displacement assumption, are summarized here for ready reference. Let (x, y, z) denote the coordinate system used to describe the deformation and stress state in a body (*e.g.*, the laminated plate). Then the strain–displacement relations (2.2–8) and the equations of motion (2.3–8) can be expressed in terms of the displacements (u, v, w), and strains, and stresses in the problem coordinates (x, y, z) as follows:

Strain–Displacement Relations:

$$\begin{aligned} \varepsilon_{xx} &= \frac{\partial u}{\partial x} \\ \varepsilon_{yy} &= \frac{\partial v}{\partial y} \\ \varepsilon_{zz} &= \frac{\partial w}{\partial z} \\ 2\varepsilon_{xy} &= \frac{\partial u}{\partial y} + \frac{\partial v}{\partial x} \\ 2\varepsilon_{xz} &= \frac{\partial u}{\partial z} + \frac{\partial w}{\partial x} \\ 2\varepsilon_{yz} &= \frac{\partial v}{\partial z} + \frac{\partial w}{\partial y} \end{aligned} \tag{2.5 – 1}$$

Equations of Motion:

$$\frac{\partial \sigma_{xx}}{\partial x} + \frac{\partial \sigma_{yx}}{\partial y} + \frac{\partial \sigma_{zx}}{\partial z} + f_x = \rho \frac{\partial^2 u}{\partial t^2} \tag{2.5 – 2a}$$

$$\frac{\partial \sigma_{xy}}{\partial x} + \frac{\partial \sigma_{yy}}{\partial y} + \frac{\partial \sigma_{zy}}{\partial z} + f_y = \rho \frac{\partial^2 v}{\partial t^2} \tag{2.5 – 2b}$$

$$\frac{\partial \sigma_{xz}}{\partial x} + \frac{\partial \sigma_{yz}}{\partial y} + \frac{\partial \sigma_{zz}}{\partial z} + f_z = \rho \frac{\partial^2 w}{\partial t^2} \tag{2.5 – 2c}$$

EXERCISE PROBLEMS

Section 2.2

2.1 Write the position of an arbitrary point $(x_1, x_2, x_3 = X_3)$ in the deformed configuration in terms of its coordinates in the undeformed configuration (shown in dotted lines) for the body shown in Figure P1. The line $X_1 = 0$ deformed to a curve of the form, $x_2 = c_1 + c_2(X_1)^2$, where c_1 and c_2 are constants.

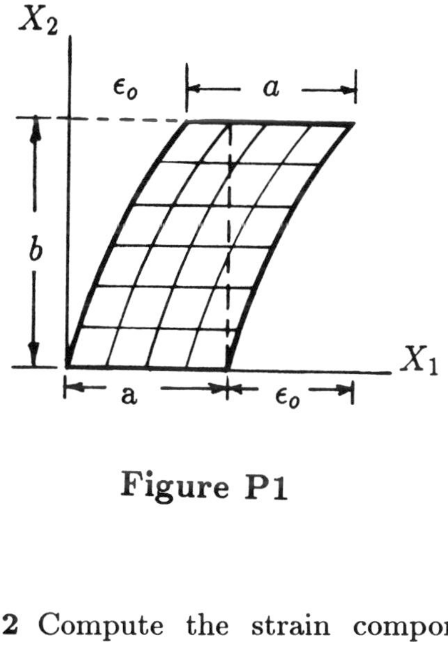

Figure P1

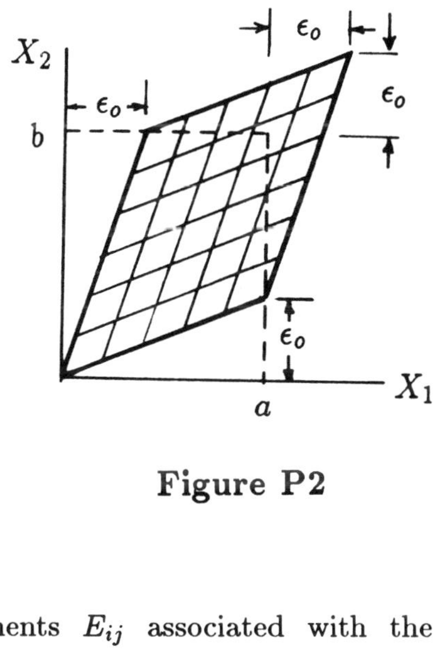

Figure P2

2.2 Compute the strain components E_{ij} associated with the displacement field

$$u_1 = \frac{\varepsilon_0}{b} X_2 \ , \ u_2 = \frac{\varepsilon_o}{a} X_1 \ , \ u_3 = 0$$

where ε_0, a and b are constants (see Figure P2).

2.3 Find the linear strains associated with the 2–D displacement field

$$u_1 = -\frac{Px_1^2 x_2}{2EI} + \frac{PL^2}{2EL} x_2 + \frac{(2+\nu)P}{6EI} x_2^3$$

$$u_2 = \frac{Ph^2(1+\nu)}{EI}(L - x_1) + \frac{\nu P}{2EI}x_1x_2^2 + \frac{P}{6EI}x_1^3 - \frac{PL^2}{2EI}x_1 + \frac{PL^3}{3EI}$$

where P, h, ν, E, I, and L are constants.

2.4 Find the linear strains associated with the 2–D displacement field

$$\begin{aligned} u_1 &= -\frac{M_o}{EI}x_1x_2 + c_1x_2 + c_2x_3 + c_4 \\ u_2 &= \frac{M_o}{2EI}\left[\nu\left(x_2^2 - x_1^2\right) + x_1^2\right] + c_4x_3 + c_5x_1 + c_6 \\ u_3 &= 0 \end{aligned}$$

where M_0, ν, E, I, and $c_1 - -c_6$ are constants.

2.5 Use the definition (2.2-7a) and the vector form of the displacement field and the del operator in the cylindrical coordinate system,

$$\begin{aligned} \mathbf{u} &= u_r\hat{\mathbf{e}}_r + u_\theta\hat{\mathbf{e}}_\theta + u_z\hat{\mathbf{e}}_z \\ \nabla &= \hat{\mathbf{e}}_r\frac{\partial}{\partial r} + \hat{\mathbf{e}}\frac{1}{r}\frac{\partial}{\partial \theta} + \hat{\mathbf{e}}_z\frac{\partial}{\partial z} \end{aligned}$$

to compute the strain–displacement relations in the cylindrical coordinate system.

2.6 Repeat Problem 2.5 for the spherical coordinate system.

2.7 If the rotation components ω_{ij} are defined by

$$\omega_{ij} = \frac{1}{2}\left(\frac{\partial u_i}{\partial x_j} - \frac{\partial u_j}{\partial x_i}\right)$$

show that

$$\omega_{ij} = \frac{1}{2}\epsilon_{ijk}\epsilon_{mnk}\frac{\partial u_m}{\partial x_n}$$

2.8 Show that in order to have a legitimate displacement corresponding to a given infinitesimal strain tensor ε, it must satisfy the compatibility relation

$$\nabla \times (\nabla \times \varepsilon)^T = \mathbf{0} \quad \text{or} \quad \epsilon_{imp}\epsilon_{jnq}\,\varepsilon_{ij,mn} = 0$$

Hint: Begin with the requirement $u_{i,jk} = u_{i,kj}$ and use Probem 2.7 results.

Section 2.3

2.9 Use the del operator and the dyadic form of σ in the cylindrical coordinate system to write the equations of motion (2.3–4) in the cylindrical coordinate system.

2.10 Repeat Problem 2.9 for the spherical coordinate system.

REFERENCES FOR ADDITIONAL READING

1. Eringen, A. C., *Mechanics of Continua*, John Wiley & Sons, New York, 1967.

2. Fung, Y. C., *Foundations of Solid Mechanics*, Prentice–Hall, Englewood Cliffs, N.J., 1965.

3. Malvern, L. E., *Introduction to the Mechanics of a Continuous Medium*, Prentice–Hall, Englewood Cliffs, N.J., 1969.

4. Lekhnitskii, S. G., *Theory of Elasticity of an Anisotropic Elastic Body*, Holden–Day, San Francisco, 1963.

5. Reddy, J. N., *Energy and Variational Methods in Applied Mechanics*, John Wiley & Sons, New York, 1984.

6. Reddy, J. N., *Mechanics of Composite Materials: Theory and Analysis,* John Wiley & Sons, New York, 1996, in press.

Chapter Three

MECHANICAL BEHAVIOR OF COMPOSITE MATERIALS

3.1 TERMINOLOGY AND DEFINITIONS

Composite materials are those formed by combining two or more materials on a macroscopic scale such that they have better engineering properties than the conventional materials, for example, metals. Some of the properties that can be improved by forming a composite material are stiffness, strength, weight, corrosion resistance, thermal properties, fatigue life and wear resistance. Most man–made composite materials are composed of two materials: a reinforcement material called fiber and a base material, called *matrix* material.

Composite materials are commonly formed in three different types: (1) *fibrous composites*, which consist of fibers of one material in a matrix material of another; (2) *particulate composites*, which are composed of particles (of macro size) of one material in a matrix of another; and (3) *laminated composites*, which are made of layers of different materials, including composites of the first two types.

The stiffness and strength of fibrous composites comes from fibers that are stiffer and stronger than the same material in bulk form. Shorter fibers, called whiskers, exhibit better strength and stiffness properties than long fibers. The matrix material keeps the fibers together, acts as a load–transfer medium between fibers, and protects fibers from being exposed to environment. Matrix materials have their usual bulk–form properties.

The particles and matrix in particulate composites can be either metallic or nonmetallic. Thus there exist four possible combinations: metallic in nonmetallic, nonmetallic in metallic, nonmetallic in nonmetallic, and metallic in metallic. Metal matrix composites are an example of nonmetallic in metallic composites.

A *lamina* or *ply* is a typical sheet of composite material. It represents a fundamental building block for composite laminates. A lamina may consist of fibers, matrix, coupling (chemical) agents and fillers. The fibers can be continuous or discontinuous, woven, unidirectional, bidirectional or randomly distributed (see Fig. 3.1–1). Unidirectional fiber–reinforced laminae exhibit the highest strength and modulus in the direction of the fibers, but they have very low strength and modulus in the transverse direction to the fibers. Discontinuous fiber–reinforced composites have lower strength and modulus than continuous fiber–reinforced composites.

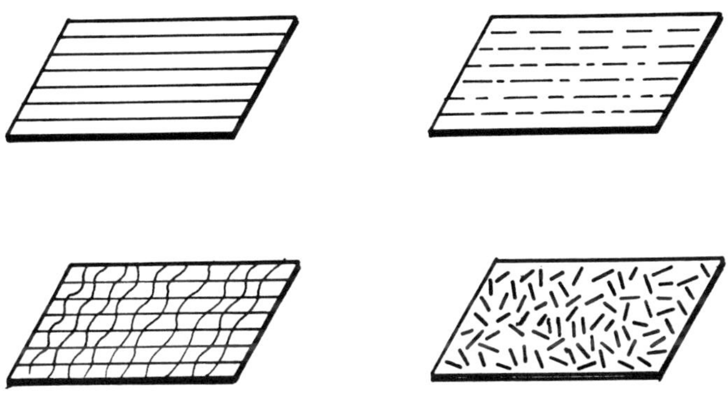

Figure 3.1–1: Fiber–reinforced composite laminae.

A *laminate* is a collection of laminae stacked to achieve the desired stiffness. For example, unidirectional fiber–reinforced laminae can be stacked so that the fibers in each lamina are oriented in the same or different directions (see Fig. 3.1–2). The sequence of various orientations of a fiber–reinforced composite layer in a laminate is termed the *lamination scheme* or *stacking sequence.* The layers are usually bonded together with the same matrix material as that in a laminae.

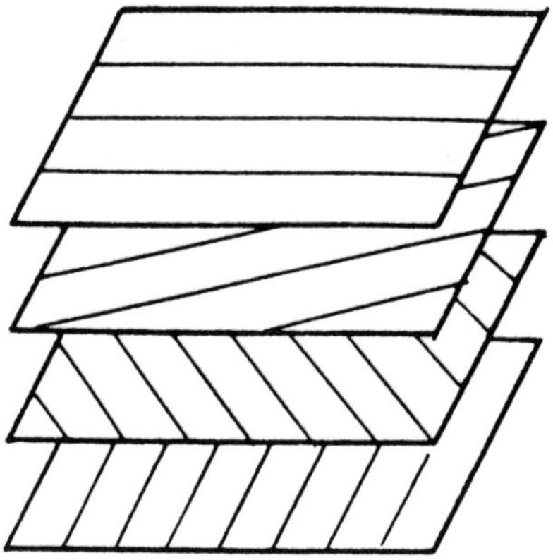

Figure 3.1–2: A typical laminate.

The lamination scheme and material properties of individual lamina provide an added flexibility to designers to tailor the stiffness and strength of the laminate to match the structural requirements. However, because of the mismatch of material properties between layers, the shear stresses produced between the layers, especially at the edges of a laminate, may cause delamination. Also, during manufacturing of laminates, material defects such as interlaminar voids, delamination, incorrect orientation, damaged fibers, and variation in thickness may be introduced. Therefore, analysis and design procedures should account for any defects.

The objective of this chapter is to study the mechanical behavior of a single lamina, treating it as a linear, orthotropic, elastic continuum. The generalized Hooke's law is revisited for an orthotropic material, the elastic coefficients of an orthotropic material are expressed in terms of engineering constants of a lamina, and the fiber–matrix interactions in a unidirectional lamina are discussed. Transformation of stresses, strains and elasticity coefficients from the lamina material coordinates to the problem coordinates are also derived.

3.2 LAMINA CONSTITUTIVE EQUATIONS

3.2–1 Generalized Hooke's Law

A material body is said to be *homogeneous* if the material properties are the same throughout the body. In a *heterogeneous* body, the material properties are a function of position. An *anisotropic* body is one that has different values of a material property in different directions at a point, *i.e.*, material properties are directional dependent (see Sec. 2.4). An *isotropic* body is one for which a material property in all directions at a point is the same. An isotropic or anisotropic material could be nonhomogeneous or homogeneous. Composite materials are inherently heterogeneous from a microscopic point of view. From a macroscopic point of view, wherein the material properties of a composite are derived from a weighted average of the constituent materials, composites are assumed to be homogeneous.

In this section we study the mechanical behavior of a typical lamina, which is the basic building block for a laminate. In formulating the constitutive equations of a lamina we assume that a lamina (a) is a continuum, and (b) it behaves as a linear elastic material. The first assumption amounts to considering the *macromechanical behavior* of a lamina. If fiber–matrix debonding and fiber breakage, for example, are to be included in the formulation of the constitutive equations of a lamina, then we must consider the *micromechanics* approach, which treats the constituent materials as continua and accounts for the mechanical behavior of the constituents and possibly their interactions. The second assumption implies that the generalized Hooke's law is valid. It should be noted that both the assumptions can be removed if we were to develop micromechanical constitutive models for inelastic (*e.g.*, plastic, viscoelastic, etc.) behavior of a lamina.

The generalized Hooke's law for an anisotropic material under isothermal conditions is (see Section 2.4) given in contracted notation by

$$\sigma_i = C_{ij}\varepsilon_j \tag{3.2 - 1}$$

where σ_{ij} are the stress components, ε_{ij} are the strain components, and C_{ij} are the material coefficients, all referred to an orthogonal Cartesian coordinate system (x_1, x_2, x_3) fixed in the body. The contracted notation is summarized in Table 3.2–1.

Table 3.2–1 Correspondence between the tensor and contracted components of stress and strain.

Stresses		Strains	
Tensor Notation	Contracted Notation	Tensor Notation	Contracted Notation
σ_{11}	σ_1	ε_{11}	ε_1
σ_{22}	σ_2	ε_{22}	ε_2
σ_{33}	σ_3	ε_{33}	ε_3
σ_{23}	σ_4	$2\varepsilon_{23}$	ε_4
σ_{13}	σ_5	$2\varepsilon_{13}$	ε_5
σ_{12}	σ_6	$2\varepsilon_{12}$	ε_6

In general, there are 36 independent constants C_{ij}. The number reduces to 21 when the material is assumed to be an hyperelastic material, *i.e.*, there exists a potential function W such that

$$\sigma_i = \frac{\partial W}{\partial \varepsilon_i} \tag{3.2 - 2}$$

When materials possess one or more planes of material symmetry, the number of independent elastic coefficients can be reduced, as discussed in Section 2.4. If a material has one plane of material symmetry (*i.e.*, monoclinic material), say about the x_1x_2–plane, then the material constants in the constitutive equations will be unchanged with respect to any other coordinate system in which the x_1x_2–plane remains unchanged. For this case the number of nonzero independent constants in the material matrix $[C]$ of a material with one plane of symmetry is 13 ($C_{ij} = C_{ji}$):

$$[C] = \begin{bmatrix} C_{11} & C_{12} & C_{13} & 0 & 0 & C_{16} \\ C_{12} & C_{22} & C_{23} & 0 & 0 & C_{26} \\ C_{13} & C_{23} & C_{33} & 0 & 0 & C_{36} \\ 0 & 0 & 0 & C_{44} & C_{45} & 0 \\ 0 & 0 & 0 & C_{45} & C_{55} & 0 \\ C_{16} & C_{26} & C_{36} & 0 & 0 & C_{66} \end{bmatrix} \qquad (3.2-3)$$

Note that, for monoclinic materials, a pure inplane shear strain gives rise to normal stress and vice versa. Hence, the principal axes of stress in general do not coincide with those of strain.

If a material has two orthogonal planes of material symmetry, material symmetry of the third mutually orthogonal plane will exist. A material with three mutually orthogonal planes of material symmetry is called *orthotropic*. If the coordinate planes are chosen parallel to the three orthogonal planes of symmetry, the stress–strain relations can be written as

$$\begin{Bmatrix} \sigma_1 \\ \sigma_2 \\ \sigma_3 \\ \sigma_4 \\ \sigma_5 \\ \sigma_6 \end{Bmatrix} = \begin{bmatrix} C_{11} & C_{12} & C_{13} & 0 & 0 & 0 \\ C_{12} & C_{22} & C_{23} & 0 & 0 & 0 \\ C_{13} & C_{23} & C_{33} & 0 & 0 & 0 \\ 0 & 0 & 0 & C_{44} & 0 & 0 \\ 0 & 0 & 0 & 0 & C_{55} & 0 \\ 0 & 0 & 0 & 0 & 0 & C_{66} \end{bmatrix} \begin{Bmatrix} \varepsilon_1 \\ \varepsilon_2 \\ \varepsilon_3 \\ \varepsilon_4 \\ \varepsilon_5 \\ \varepsilon_6 \end{Bmatrix} \qquad (3.2-4)$$

The inverse relations, strain–stress relations, are given by

$$\begin{Bmatrix} \varepsilon_1 \\ \varepsilon_2 \\ \varepsilon_3 \\ \varepsilon_4 \\ \varepsilon_5 \\ \varepsilon_6 \end{Bmatrix} = \begin{bmatrix} S_{11} & S_{12} & S_{13} & 0 & 0 & 0 \\ S_{12} & S_{22} & S_{23} & 0 & 0 & 0 \\ S_{13} & S_{23} & S_{33} & 0 & 0 & 0 \\ 0 & 0 & 0 & S_{44} & 0 & 0 \\ 0 & 0 & 0 & 0 & S_{55} & 0 \\ 0 & 0 & 0 & 0 & 0 & S_{66} \end{bmatrix} \begin{Bmatrix} \sigma_1 \\ \sigma_2 \\ \sigma_3 \\ \sigma_4 \\ \sigma_5 \\ \sigma_6 \end{Bmatrix} \qquad (3.2-5)$$

where S_{ij} are called the *material compliance coefficients.*

If at every point of a material there is one plane in which the material properties are equal in all directions, then the material is called *transversely isotropic.* For transversely isotropic material there are five independent constants. The stress–strain relations for a transversely isotropic material with x_1x_2-plane as the special plane of isotropy are

$$\begin{Bmatrix} \sigma_1 \\ \sigma_2 \\ \sigma_3 \\ \sigma_4 \\ \sigma_5 \\ \sigma_6 \end{Bmatrix} = \begin{bmatrix} C_{11} & C_{12} & C_{13} & 0 & 0 & 0 \\ C_{12} & C_{11} & C_{13} & 0 & 0 & 0 \\ C_{13} & C_{13} & C_{33} & 0 & 0 & 0 \\ 0 & 0 & 0 & C_{44} & 0 & 0 \\ 0 & 0 & 0 & 0 & C_{44} & 0 \\ 0 & 0 & 0 & 0 & 0 & C \end{bmatrix} \begin{Bmatrix} \varepsilon_1 \\ \varepsilon_2 \\ \varepsilon_3 \\ \varepsilon_4 \\ \varepsilon_5 \\ \varepsilon_6 \end{Bmatrix} \qquad (3.2-6)$$

where $C = (C_{11} - C_{12})/2$.

An *isotropic* material is one for which the material properties are independent of the direction, *i.e.*, there are an infinite number of planes of material symmetry. For isotropic materials there are only two independent elastic constants, irrespective of whether the strain energy function exists or not. The stress–strain relations for an isotropic material are given by

$$\begin{Bmatrix} \sigma_1 \\ \sigma_2 \\ \sigma_3 \\ \sigma_4 \\ \sigma_5 \\ \sigma_6 \end{Bmatrix} = \begin{bmatrix} C_{11} & C_{12} & C_{12} & 0 & 0 & 0 \\ C_{12} & C_{11} & C_{12} & 0 & 0 & 0 \\ C_{12} & C_{12} & C_{11} & 0 & 0 & 0 \\ 0 & 0 & 0 & C & 0 & 0 \\ 0 & 0 & 0 & 0 & C & 0 \\ 0 & 0 & 0 & 0 & 0 & C \end{bmatrix} \begin{Bmatrix} \varepsilon_1 \\ \varepsilon_2 \\ \varepsilon_3 \\ \varepsilon_4 \\ \varepsilon_5 \\ \varepsilon_6 \end{Bmatrix} \qquad (3.2-7)$$

where $C = (C_{11} - C_{12})/2$.

The material coefficients in the constitutive equation are assumed to be constant during deformation. When temperature changes occurs in an elastic body, we must account for the thermal expansion of the material, even though we neglect the variation of elastic constants with temperature. When the strains, geometric changes and temperature variations are sufficiently small, all governing equations are linear, and superposition of mechanical and thermal effects is allowed. The linear thermoelastic constitutive equations have the form,

$$\sigma_j = C_{ji}[-\alpha_i(T - T_o) + \varepsilon_i] \qquad (3.2-8)$$

$$\varepsilon_j = S_{ji}\sigma_i + \alpha_i(T - T_o) \qquad (3.2-9)$$

where $T - T_o$ denotes the temperature change from reference temperature T_o, α_i denote the linear coefficients of thermal expansion, and summation on repeated subscripts is taken in the range of 1 to 6. For orthotropic materials α_4, α_5 and α_6 are zero, and for isotropic behavior we have $\alpha_1 = \alpha_2 = \alpha_3$.

3.2–2 Engineering Constants of Orthotropic Materials

The constitutive equations for a material that has three orthogonal planes of material property symmetry (*i.e.*, orthotropic materials), referred to the principal material directions are given by Eq. (4). The material coefficients C_{ij} and S_{ij} can be expressed in terms of engineering constants, *i.e.*, principal moduli, Poisson's ratios, and shear moduli:

$$C_{11} = E_1 \cdot \frac{1 - \nu_{23}\nu_{32}}{\Delta}, \quad C_{12} = E_1 \cdot \frac{\nu_{21} + \nu_{31}\nu_{23}}{\Delta} = E_2 \cdot \frac{\nu_{12} + \nu_{32}\nu_{13}}{\Delta}$$

$$C_{13} = E_1 \cdot \frac{\nu_{31} + \nu_{21}\nu_{32}}{\Delta} = E_3 \cdot \frac{\nu_{13} + \nu_{12}\nu_{23}}{\Delta}, \quad C_{22} = E_2 \cdot \frac{1 - \nu_{13}\nu_{31}}{\Delta}$$

$$C_{23} = E_2 \cdot \frac{\nu_{32} + \nu_{12}\nu_{31}}{\Delta} = E_3 \cdot \frac{\nu_{23} + \nu_{21}\nu_{13}}{\Delta}, \quad C_{33} = E_3 \cdot \frac{1 - \nu_{12}\nu_{21}}{\Delta\Delta}$$

$$\Delta = 1 - \nu_{12}\nu_{21} - \nu_{23}\nu_{32} - \nu_{31}\nu_{13} - 2\nu_{21}\nu_{32}\nu_{13} \qquad (3.2-10a)$$

where

E_1, E_2, E_3 = Young's moduli in 1, 2 and 3 directions, respectively,

ν_{ij} = Poisson's ratio — defined as the ratio of transverse strain in the j-direction to the axial strain in the i-direction, when stressed in the i-direction $-\varepsilon_j/\varepsilon_i$ (for $\sigma_i = \sigma$ and all other stresses are zero), $i, j = 1, 2, 3$, and

G_{23}, G_{13}, G_{12} = shear moduli in the 2-3, 1-3, and 1-2 planes, respectively.

$$(3.2-10b)$$

Because of their direct and obvious physical meaning, engineering constants are used in place of the more abstract stiffness and compliance coefficients of Eqs. (4) and (5).

The strain–stress relations (5) for orthotropic materials can be expressed in terms of engineering constants as

$$\begin{Bmatrix} \varepsilon_1 \\ \varepsilon_2 \\ \varepsilon_3 \\ \varepsilon_4 \\ \varepsilon_5 \\ \varepsilon_6 \end{Bmatrix} = \begin{bmatrix} \frac{1}{E_1} & -\frac{\nu_{21}}{E_2} & -\frac{\nu_{31}}{E_2} & 0 & 0 & 0 \\ -\frac{\nu_{12}}{E_1} & \frac{1}{E_2} & -\frac{\nu_{32}}{E_3} & 0 & 0 & 0 \\ -\frac{\nu_{13}}{E_1} & -\frac{\nu_{23}}{E_2} & \frac{1}{E_3} & 0 & 0 & 0 \\ 0 & 0 & 0 & \frac{1}{G_{23}} & 0 & 0 \\ 0 & 0 & 0 & 0 & \frac{1}{G_{31}} & 0 \\ 0 & 0 & 0 & 0 & 0 & \frac{1}{G_{12}} \end{bmatrix} \begin{Bmatrix} \sigma_1 \\ \sigma_2 \\ \sigma_3 \\ \sigma_4 \\ \sigma_5 \\ \sigma_6 \end{Bmatrix} \qquad (3.2-11)$$

Since the compliance matrix $[S]$ is the inverse of the stiffness matrix $[C]$, which is symmetric, the compliance matrix $[S]$ is a symmetric matrix. This implies, by inspection of Eq. (11), that the following reciprocal relations hold:

$$\frac{\nu_{21}}{E_2} = \frac{\nu_{12}}{E_1}, \quad \frac{\nu_{31}}{E_3} = \frac{\nu_{13}}{E_1}, \quad \frac{\nu_{32}}{E_3} = \frac{\nu_{23}}{E_2} \qquad (3.2-12a)$$

or, in short

$$\frac{\nu_{ij}}{E_i} = \frac{\nu_{ji}}{E_j} \quad \begin{matrix} (\text{no sum on } i \text{ and } j) \\ i, j = 1, 2, 3 \end{matrix} \qquad (3.2-12b)$$

Thus, there are only nine independent engineering constants

$$E_1, E_2, E_3, G_{23}, G_{31}, G_{12}, \nu_{12}, \nu_{31}, \nu_{23}$$

for an orthotropic material, since ν_{21}, ν_{31} and ν_{32} can be expressed in terms of the three principal moduli (E_1, E_2, E_3) and Poisson's ratios $(\nu_{12}, \nu_{13}, \nu_{23})$.

Note that

$$\nu_{12}\nu_{23}\nu_{31} = \left(\frac{E_1}{E_2}\right)\nu_{21}\left(\frac{E_2}{E_3}\right)\nu_{32}\left(\frac{E_3}{E_1}\right)\nu_{13} = \nu_{21}\nu_{32}\nu_{13} \quad (3.2-13)$$

Physical arguments can be used to restrict Young's moduli and shear moduli to be positive:

$$E_1, E_2, E_3, G_{12}, G_{23}, G_{31} > 0$$

For isotropic materials we have

$$E_1 = E_2 = E_3 = E \ , \ G_{12} = G_{13} = G_{13} = G_{32} \equiv G$$

$$\nu_{12} = \nu_{23} = \nu_{13} \equiv \nu$$

Of the three constants (E, ν, G), only two are linearly independent, and the third one is related to the other two by the relation

$$G = \frac{E}{2(1+\nu)} \quad (3.2-14)$$

3.2–3 Characterization of a Unidirectional Lamina

A unidirectional fiber–reinforced lamina is treated as an orthotropic material whose material symmetry planes are parallel and transverse to the fiber direction. The material coordinate axis x_1 is taken to be parallel to the fiber, the x_2–axis transverse to the fiber direction in the plane of the lamina, and the x_3–axis is perpendicular to the plane of the lamina (see Fig. 3.2–1). The orthotropic material properties of a lamina are obtained either by the theoretical approach or by laboratory testing.

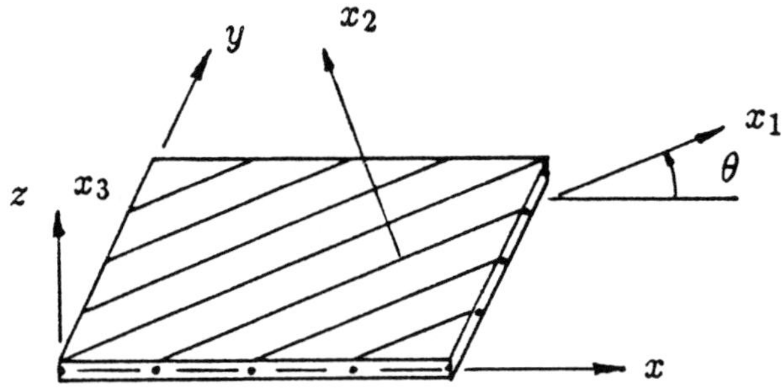

Figure 3.2–1: Global and material coordinates of a lamina.

The theoretical approach, called a *micromechanics approach*, used to determine the engineering constants of a continuous fiber–reinforced composite material is based on the following assumptions:

1. Perfect bonding between fibers and matrix exists.
2. Fibers are parallel and uniformly distributed throughout.
3. The matrix is free of voids or microcracks and initially in a stress–free state.
4. Both fibers and matrix are isotropic and obey Hooke's law.
5. The applied loads are either parallel or perpendicular to the fiber direction.

Let

E_f = modulus of the fiber

E_m = modulus of the matrix

ν_f = Poisson's ratio of the fiber

ν_m = Poisson's ratio of the matrix

V_f = fiber volume fraction

V_m = matrix volume fraction

Then it can be shown that the lamina engineering constants are given by

$$E_1 = \text{longitudinal modulus}$$

$$= E_f V_f + E_m V_m \qquad (3.2-14a)$$

$$E_2 = \text{Transverse modulus}$$

$$= \frac{E_f E_m}{E_f V_m + E_m V_f} \qquad (3.2-14b)$$

ν_{12} = Major Poisson's ratio, and

$$G_{12} = \text{Shear modulus}$$

$$= \frac{G_f G_m}{G_f V_m + G_m V_f} \qquad (3.2-15a)$$

where

$$G_f = \frac{E_f}{2(1+\nu_f)} \quad \text{and} \quad G_m = \frac{E_m}{2(1+\nu_m)} \qquad (3.2-15b)$$

The engineering constants E_1, E_2, E_3, G_{12}, G_{13}, G_{23}, ν_{12}, ν_{13} and ν_{23} of an orthotropic material are determined experimentally using an appropriate test specimen made of the material. For example, the values of E_1 and ν_{12} of a fiber–reinforced material are measured using a uniaxial test specimen. The specimen consists of several layers of the material with fibers in each layer being aligned with the longitudinal direction. The specimen is then loaded along the longitudinal direction and strains along and perpendicular to the fiber directions are measured using strain gages. If the load applied is P, the stress along the fiber direction is $\sigma_1 = P/A$, where A denotes the cross–sectional area of the specimen. For different values of the load P, σ_1 and longitudinal and lateral strains ε_1 and ε_2 can be recorded. Then E_1 denotes the slope of the experimental relation between σ_1 and ε_1 whereas ν_{12} denotes the negative of the slope of the experimental relation between the lateral strain ε_2 and longitudinal strain ε_1. Similarly, other constants can be determined from appropriate tests.

3.3 TRANSFORMATION OF STRESSES AND STRAINS

3.3–1 Coordinate Transformations

In the preceding sections we discussed constitutive equations for a hyperelastic material. The equations for monoclinic, orthotropic and transversely isotropic materials were written in terms of the stress and strain components that are referred to the coordinate system that coincides with the principal material coordinate system. The coordinate system used in the solution of a problem, in general, does not coincide with the material coordinate system. Further, composite laminates have several layers with different orientation of their material coordinates with respect to the laminate coordinates. Since all quantities should be referred to a single coordinate system, we need to establish transformation relations among stresses and strains in one coordinate system to the corresponding quantities in another coordinate system. These relations can be used to transform constitutive equations from the material coordinates to the coordinates used in the solution of a problem.

Let (x, y, z) denote the coordinates used in the analytical description of a problem, and let (x_1, x_2, x_3) be the principal material coordinates such that (x_1, x_2, x_3) is obtained from (x, y, z) by rotating the xy–plane counterclockwise by an angle θ about the z–axis (see Fig. 3.2–1). The two sets of coordinates are related by [see Eq. (1.3–18)]

$$\begin{Bmatrix} x_1 \\ x_2 \\ x_3 \end{Bmatrix} = \begin{bmatrix} \cos\theta & \sin\theta & 0 \\ -\sin\theta & \cos\theta & 0 \\ 0 & 0 & 1 \end{bmatrix} \begin{Bmatrix} x \\ y \\ z \end{Bmatrix} = [A] \begin{Bmatrix} x \\ y \\ z \end{Bmatrix} \qquad (3.3-1)$$

The inverse of Eq. (1) is

$$\begin{Bmatrix} x \\ y \\ z \end{Bmatrix} = \begin{bmatrix} \cos\theta & -\sin\theta & 0 \\ \sin\theta & \cos\theta & 0 \\ 0 & 0 & 1 \end{bmatrix} \begin{Bmatrix} x_1 \\ x_2 \\ x_3 \end{Bmatrix} = [A]^{-1} \begin{Bmatrix} x_1 \\ x_2 \\ x_3 \end{Bmatrix} \qquad (3.3-2)$$

Note that the inverse of $[A]$ is equal to its transpose: $[A]^{-1} = [A]^T$.

The transformation relation (1) and (2) are also valid for the unit vectors associated with the two coordinate systems [see Eqs. (1.3–14) and (1.3–15)]:

$$\begin{Bmatrix} \hat{\mathbf{e}}_1 \\ \hat{\mathbf{e}}_2 \\ \hat{\mathbf{e}}_3 \end{Bmatrix} = [A] \begin{Bmatrix} \hat{\mathbf{e}}_x \\ \hat{\mathbf{e}}_y \\ \hat{\mathbf{e}}_z \end{Bmatrix}, \quad \begin{Bmatrix} \hat{\mathbf{e}}_x \\ \hat{\mathbf{e}}_y \\ \hat{\mathbf{e}}_z \end{Bmatrix} = [A]^T \begin{Bmatrix} \hat{\mathbf{e}}_1 \\ \hat{\mathbf{e}}_2 \\ \hat{\mathbf{e}}_3 \end{Bmatrix} \qquad (3.3-3)$$

The coordinate transformations in Eqs. (1) and (2) are useful in transforming vector and tensorial quantities from the principal material coordinates to global coordinates. In the next section we consider the transformation of stress components from material coordinates to global (or problem) coordinates.

3.3–2 Transformation of Stress Components

Since stress is a second–order tensor, there are two basis vectors associated with it that must be transformed using equations of the form (3). We have from Eq. (1.3–23) the transformation relation

$$[\sigma] = [A]^T[\bar{\sigma}][A] \qquad (3.3-4)$$

where $[A]$ is defined in Eq. (1), and σ_{ij} are the components of the stress tensor in the system $(x = x_1, y = x_2, z = x_3)$, whereas $\bar{\sigma}_{ij}$ are the components of the same stress tensor in the system $(x_1 = \bar{x}_1, x_2 = \bar{x}_2, x_3 = \bar{x}_3)$, and

$$\sigma_{11} = \sigma_{xx} \;, \quad \sigma_{12} = \sigma_{xy} \;, \quad \cdots \quad , \sigma_{33} = \sigma_{zz} \tag{3.3 - 5}$$

Carrying out the matrix multiplications in Eq. (4), and writing the result in a single–column format we have

$$\begin{Bmatrix} \sigma_{xx} \\ \sigma_{yy} \\ \sigma_{zz} \\ \sigma_{yz} \\ \sigma_{xz} \\ \sigma_{xy} \end{Bmatrix} = [T_1] \begin{Bmatrix} \sigma_1 \\ \sigma_2 \\ \sigma_3 \\ \sigma_4 \\ \sigma_5 \\ \sigma_6 \end{Bmatrix} \tag{3.3 - 6a}$$

or

$$\{\sigma\} = [T_1]\{\bar{\sigma}\} \tag{3.3 - 6b}$$

where

$$[T_1] = \begin{bmatrix} \cos^2\theta & \sin^2\theta & 0 & 0 & 0 & -2\sin\theta\cos\theta \\ \sin^2\theta & \cos^2\theta & 0 & 0 & 0 & 2\sin\theta\cos\theta \\ 0 & 0 & 1 & 0 & 0 & 0 \\ 0 & 0 & 0 & \cos\theta & \sin\theta & 0 \\ 0 & 0 & 0 & -\sin\theta & \cos\theta & 0 \\ \sin\theta\cos\theta & -\sin\theta\cos\theta & 0 & 0 & 0 & \cos^2\theta - \sin^2\theta \end{bmatrix} \tag{3.3 - 7}$$

The inverse relationship between σ_{ij} and $\bar{\sigma}_{ij}$ is given by

$$\{\bar{\sigma}\} = [T_2]\{\sigma\} \tag{3.3 - 8a}$$

where

$$[T_2] = \begin{bmatrix} \cos^2\theta & \sin^2\theta & 0 & 0 & 0 & 2\sin\theta\cos\theta \\ \sin^2\theta & \cos^2\theta & 0 & 0 & 0 & -2\sin\theta\cos\theta \\ 0 & 0 & 1 & 0 & 0 & 0 \\ 0 & 0 & 0 & \cos\theta & -\sin\theta & 0 \\ 0 & 0 & 0 & \sin\theta & \cos\theta & 0 \\ -\sin\theta\cos\theta & \sin\theta\cos\theta & 0 & 0 & 0 & \cos^2\theta - \sin^2\theta \end{bmatrix} \quad (3.3-8b)$$

3.3–3 Transformation of Strain Components

Similar transformation equations can be written for the engineering strains. For tensor components of strains ε_{ij} and $\bar{\varepsilon}_{ij}$ in the two coordinate systems, Eqs. (7) and (10) are valid with $\bar{\sigma}_{ij}$ replaced by $\bar{\varepsilon}_{ij}$ and σ_{ij} replaced by ε_{ij}. However, the single–column formats for strains are slightly different because of the following definition of the engineering shear strains in terms of the tensor components of shear strain:

$$2\varepsilon_{12} = \varepsilon_6, \ 2\varepsilon_{13} = \varepsilon_5, \ 2\varepsilon_{23} = \varepsilon_4 \quad (3.3-9a)$$

$$2\varepsilon_{xy} = \gamma_{xy}, \ 2\varepsilon_{xz} = \gamma_{xz}, \ 2\varepsilon_{yz} = \gamma_{xz} \quad (3.3-9b)$$

We have

$$\begin{Bmatrix} \varepsilon_{xx} \\ \varepsilon_{yy} \\ \varepsilon_{zz} \\ \gamma_{yz} \\ \gamma_{xz} \\ \gamma_{xy} \end{Bmatrix} = [R_1] \begin{Bmatrix} \varepsilon_1 \\ \varepsilon_2 \\ \varepsilon_3 \\ \varepsilon_4 \\ \varepsilon_5 \\ \varepsilon_6 \end{Bmatrix} \quad (3.3-10a)$$

where

$$[R_1] = \begin{bmatrix} \cos^2\theta & \sin^2\theta & 0 & 0 & 0 & -\sin\theta\cos\theta \\ \sin^2\theta & \cos^2\theta & 0 & 0 & 0 & \sin\theta\cos\theta \\ 0 & 0 & 1 & 0 & 0 & 0 \\ 0 & 0 & 0 & \cos\theta & \sin\theta & 0 \\ 0 & 0 & 0 & -\sin\theta & \cos\theta & 0 \\ 2\sin\theta\cos\theta & -2\sin\theta\cos\theta & 0 & 0 & 0 & \cos^2\theta - \sin^2\theta \end{bmatrix} \tag{3.3-10b}$$

The inverse relation is given by

$$\begin{Bmatrix} \varepsilon_1 \\ \varepsilon_2 \\ \varepsilon_3 \\ \varepsilon_4 \\ \varepsilon_5 \\ \varepsilon_6 \end{Bmatrix} = [R_2] \begin{Bmatrix} \varepsilon_{xx} \\ \varepsilon_{yy} \\ \varepsilon_{zz} \\ \gamma_{yz} \\ \gamma_{xz} \\ \gamma_{xy} \end{Bmatrix} \tag{3.3-11a}$$

where

$$[R_2] = \begin{bmatrix} \cos^2\theta & \sin^2\theta & 0 & 0 & 0 & \sin\theta\cos\theta \\ \sin^2\theta & \cos^2\theta & 0 & 0 & 0 & -\sin\theta\cos\theta \\ 0 & 0 & 1 & 0 & 0 & 0 \\ 0 & 0 & 0 & \cos\theta & -\sin\theta & 0 \\ 0 & 0 & 0 & \sin\theta & \cos\theta & 0 \\ -2\sin\theta\cos\theta & 2\sin\theta\cos\theta & 0 & 0 & 0 & \cos^2\theta - \sin^2\theta \end{bmatrix} \tag{3.3-11b}$$

It is interesting to note that the transformation matrix $[T_1]$ in Eq. (6b) is the transpose of the matrix $[R_2]$ in Eq. (11b). Similarly, the transformation matrix $[R_1]$ in Eq. (10b) is the transpose of the matrix $[T_2]$ in Eq. (8b):

$$\{\bar{\varepsilon}\} = [R_1]\{\varepsilon\} = [T_2]^T\{\varepsilon\} \tag{3.3-12a}$$

$$\{\varepsilon\} = [R_2]\{\varepsilon\} = [T_1]^T\{\bar{\varepsilon}\} \tag{3.3-12b}$$

Equations (6)–(12) are useful in transforming the stress–strain relations from material (or local) coordinates (x_1, x_2, x_3) to those in problem (or global) coordinates (x, y, z).

3.3–4 Transformation of Elastic Coefficients

The components of the fourth–order elasticity tensor $\bar{C}_{ijk\ell}$ in the global coordinate system can be related to the components C_{mnpq} in the material coordinates. The transformation relations can be obtained using Eq. (6b), the Hooke's law in the material coordinates, and Eq. (12b):

$$\{\sigma\} = [T^1]\{\bar{\sigma}\} = [T^1][C]\{\bar{\varepsilon}\} = [T^1][C][T^1]^T\{\varepsilon\} \qquad (3.3-13a)$$

where $[C]$ denotes the elastic stiffnesses in the material coordinates. However, by Hooke's law in the global coordinate system we have

$$\{\sigma\} = [\bar{C}]\{\varepsilon\} \qquad (3.3-13b)$$

where $[\bar{C}]$ denotes the elastic stiffnesses in the global coordinates. Hence, it follows from Eqs. (13a) and (13b) that

$$[\bar{C}] = [T^1][C][T^1]^T \qquad (3.3-14)$$

Carrying out the matrix operations in Eq. (14), we obtain

$$\begin{aligned}
\bar{C}_{11} &= C_{11}\cos^4\theta + 2(C_{12} + 2C_{66})\sin^2\theta\ \cos^2\theta + C_{22}\sin^4\theta \\
\bar{C}_{12} &= \hat{C}_{12}\sin^2\theta\ \cos^2\theta + C_{12}(\sin^4\theta + \cos^4\theta) \\
\bar{C}_{13} &= C_{13}\cos^2\theta + C_{23}\sin^2\theta \\
\bar{C}_{16} &= \hat{C}_{16}\cos^3\theta\ \sin\theta + \hat{C}_{26}\sin^3\theta\ \cos\theta \\
\bar{C}_{22} &= C_{11}\sin^4\theta + 2(C_{12} + 2C_{66})\sin^2\theta\ \cos^2\theta + C_{22}\cos^4\theta \\
\bar{C}_{23} &= C_{13}\sin^2\theta + C_{23}\cos^2\theta \\
\bar{C}_{26} &= \hat{C}_{16}\sin^3\theta\ \cos\theta + \hat{C}_{26}\cos^3\theta\ \sin\theta \\
\bar{C}_{33} &= C_{33} \\
\bar{C}_{36} &= (C_{13} - C_{23})\cos\theta\ \sin\theta \\
\bar{C}_{44} &= C_{44}\cos^2\theta + C_{55}\sin^2\theta \\
\bar{C}_{45} &= (C_{55} - C_{44})\cos\theta\ \sin\theta \\
\bar{C}_{55} &= C_{44}\sin^2\theta + C_{55}\cos^2\theta \\
\bar{C}_{66} &= \hat{C}_{66}\sin^2\theta\ \cos^2\theta + C_{66}(\sin^4\theta + \cos^4\theta)
\end{aligned} \qquad (3.3-15)$$

where

$$\begin{aligned}
\hat{C}_{12} &= C_{11} + C_{22} - 4C_{66} \\
\hat{C}_{16} &= C_{11} - C_{12} - 2C_{66} \\
\hat{C}_{26} &= C_{12} - C_{22} + 2C_{66} \\
\hat{C}_{66} &= C_{11} + C_{22} - 2C_{12} - 2C_{66}
\end{aligned} \qquad (3.3-15b)$$

and the global stress–strain relations are given by

$$\begin{Bmatrix} \sigma_{xx} \\ \sigma_{yy} \\ \sigma_{zz} \\ \sigma_{yz} \\ \sigma_{xz} \\ \sigma_{xy} \end{Bmatrix} = \begin{bmatrix} \bar{C}_{11} & \bar{C}_{12} & \bar{C}_{13} & 0 & 0 & \bar{C}_{16} \\ \bar{C}_{21} & \bar{C}_{22} & \bar{C}_{13} & 0 & 0 & \bar{C}_{26} \\ \bar{C}_{31} & \bar{C}_{32} & \bar{C}_{33} & 0 & 0 & \bar{C}_{36} \\ 0 & 0 & 0 & \bar{C}_{44} & \bar{C}_{45} & 0 \\ 0 & 0 & 0 & \bar{C}_{54} & \bar{C}_{55} & 0 \\ \bar{C}_{16} & \bar{C}_{26} & \bar{C}_{36} & 0 & 0 & \bar{C}_{66} \end{bmatrix} \begin{Bmatrix} \varepsilon_{xx} \\ \varepsilon_{yy} \\ \varepsilon_{zz} \\ 2\varepsilon_{yz} \\ 2\varepsilon_{xz} \\ 2\varepsilon_{xy} \end{Bmatrix} \qquad (3.3-16)$$

where the $\bar{C}_{ij}$ are the transformed elastic coefficients in the (x, y, z) system, which are related to the elastic coefficients in the material coordinates, C_{ij} by Eqs. (15).

The transformation relations (15) and (16) are valid for an arbitrary coordinate system (x_1, x_2, x_3) which is oriented at an angle θ (in the xy–plane) from the (x, y, z) coordinate system (see Fig. 3.2–1). The orientation angle θ is measured counterclockwise from the x–axis to the x_1–axis. The stress–strain relations in the principal material coordinates (x_1, x_2, x_3) of an orthotropic lamina are given by Eq. (3.2–4), and the stress–strain relations of the lamina in the global coordinate system (x, y, z) are given by Eq. (16). The material coefficients of the lamina in the (x, y, z) coordinate system are related to material coefficients in the material coordinates by Eq. (15).

3.4 CONSTITUTIVE RELATIONS FOR PLANE STRESS

A plane stress state is defined to be one in which all transverse stresses are negligible. Most laminates are typically thin and experience a state of plane stress. For a lamina in the x_1x_2–plane, the following stress components are set to zero (see Fig. 3.4–1):

$$\sigma_3 = 0, \;\; \sigma_4 = 0, \;\; \sigma_5 = 0 \;\;\; (\sigma_{33} = 0, \;\; \sigma_{32} = 0, \;\; \sigma_{31} = 0) \qquad (3.4-1a)$$

For an orthotropic lamina this assumption results in the following transverse strains [see Eq. (3.2–11)]

$$\varepsilon_3 = S_{13}\sigma_1 + S_{23}\sigma_2, \quad \varepsilon_4 = 0, \ \varepsilon_5 = 0 \qquad (3.4-1b)$$

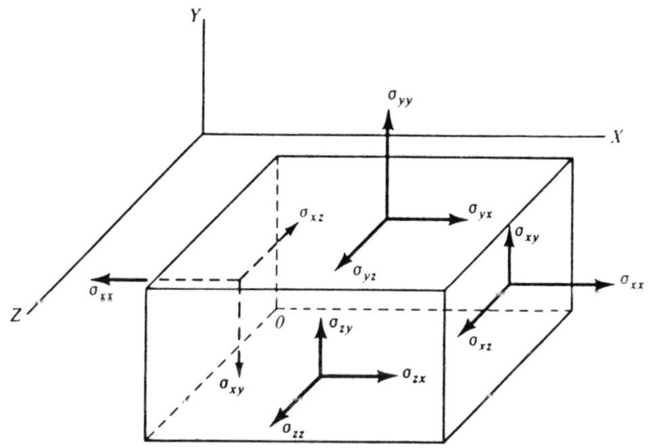

Figure 3.4–1: A plane state of stress.

The strain–stress relations (3.2–11) for the plane stress case reduce to

$$\begin{Bmatrix} \varepsilon_1 \\ \varepsilon_2 \\ \varepsilon_6 \end{Bmatrix} = \begin{bmatrix} S_{11} & S_{12} & 0 \\ S_{12} & S_{22} & 0 \\ 0 & 0 & S_{66} \end{bmatrix} \begin{Bmatrix} \sigma_1 \\ \sigma_2 \\ \sigma_6 \end{Bmatrix} = \begin{bmatrix} \frac{1}{E_1} & -\frac{\nu_{12}}{E_1} & 0 \\ -\frac{\nu_{12}}{E_1} & \frac{1}{E_1} & \\ 0 & 0 & \frac{1}{G_{12}} \end{bmatrix} \begin{Bmatrix} \sigma_1 \\ \sigma_2 \\ \sigma_6 \end{Bmatrix}$$

$$\varepsilon_3 = -\frac{\nu_{13}}{E_1}\sigma_1 - \frac{\nu_{23}}{E_2}\sigma_2 \qquad (3.4-2)$$

The strain–stress relations (2) can be inverted to obtain the stress strain relations

$$\begin{Bmatrix} \sigma_1 \\ \sigma_2 \\ \sigma_6 \end{Bmatrix} = \begin{bmatrix} Q_{11} & Q_{12} & 0 \\ Q_{12} & Q_{22} & 0 \\ 0 & 0 & Q_{66} \end{bmatrix} \begin{Bmatrix} \varepsilon_1 \\ \varepsilon_2 \\ \varepsilon_3 \end{Bmatrix} \tag{3.4 – 3a}$$

where the Q_{ij}, called *plane stress–reduced stiffnesses*, are given by

$$\begin{aligned} Q_{11} &= \frac{S_{22}}{S_{11}S_{22} - S_{12}^2} = \frac{E_1}{1 - \nu_{12}\nu_{21}} \\ Q_{12} &= \frac{S_{12}}{S_{11}S_{22} - S_{12}^2} = \frac{\nu_{12}E_2}{1 - \nu_{12}\nu_{21}} \\ Q_{22} &= \frac{S_{11}}{S_{11}S_{22} - S_{12}^2} = \frac{E_2}{1 - \nu_{12}\nu_{21}} \\ Q_{66} &= \frac{1}{S_{66}} = G_{12} \end{aligned} \tag{3.4 – 3b}$$

Note that the plane stress–reduced stiffnesses involve four independent material constants, E_1, E_2, ν_{12} and G_{12}.

The thermoelastic constitutive equations for an orthotropic lamina are

$$\begin{Bmatrix} \sigma_1 \\ \sigma_2 \\ \sigma_6 \end{Bmatrix} = \begin{bmatrix} Q_{11} & Q_{12} & 0 \\ Q_{12} & Q_{22} & 0 \\ 0 & 0 & Q_{66} \end{bmatrix} \begin{Bmatrix} \varepsilon_1 - \alpha_1 \Delta T \\ \varepsilon_2 - \alpha_2 \Delta T \\ \varepsilon_6 \end{Bmatrix} \tag{3.4 – 4}$$

where α_1 and α_2 are the coefficients of thermal expansion along the x_1 and x_2 directions, respectively.

The transformed stress–strain relations of a lamina in plane state of stress are

$$\begin{Bmatrix} \sigma_{xx} \\ \sigma_{yy} \\ \sigma_{xy} \end{Bmatrix} = \begin{bmatrix} \bar{Q}_{11} & \bar{Q}_{12} & \bar{Q}_{16} \\ \bar{Q}_{12} & \bar{Q}_{22} & \bar{Q}_{26} \\ \bar{Q}_{16} & \bar{Q}_{26} & \bar{Q}_{66} \end{bmatrix} \begin{Bmatrix} \varepsilon_{xx} - \alpha_{xx}\,\Delta T \\ \varepsilon_{yy} - \alpha_{yy}\,\Delta T \\ 2\varepsilon_{xy} - \alpha_{xy}\,\Delta T \end{Bmatrix} \tag{3.4 – 5}$$

where

$$\begin{aligned}
\bar{Q}_{11} &= Q_{11}\cos^4\theta + 2(Q_{12}+2Q_{66})\sin^2\theta\ \cos^2\theta + Q_{22}\sin^4\theta \\
\bar{Q}_{12} &= \hat{Q}_{12}\sin^2\theta\ \cos^2\theta + Q_{12}(\sin^4\theta + \cos^4\theta) \\
\bar{Q}_{22} &= Q_{11}\sin^4\theta + 2(Q_{12}+2Q_{66})\sin^2\theta\ \cos^2\theta + Q_{22}\cos^4\theta \\
\bar{Q}_{16} &= \hat{Q}_{16}\sin\theta\ \cos^3\theta + \hat{Q}_{26}\sin^3\theta\cos\theta \\
\bar{Q}_{26} &= \hat{Q}_{16}\sin^3\theta\ \cos\theta + \hat{Q}_{16}\sin\theta\ \cos^3\theta \\
\bar{Q}_{66} &= \hat{Q}_{66}\sin^2\theta\ \cos^2\theta + Q_{66}(\sin^4\theta + \cos^4\theta) \qquad (3.4-6a)
\end{aligned}$$

where

$$\begin{aligned}
\hat{Q}_{12} &= Q_{11} + Q_{22} - 4Q_{66} \\
\hat{Q}_{16} &= Q_{11} - Q_{12} - 2Q_{66} \\
\hat{Q}_{26} &= Q_{12} - Q_{22} + 2Q_{66} \\
\hat{Q}_{66} &= Q_{11} + Q_{22} - 2Q_{12} - 2Q_{66} \qquad (3.4-6b)
\end{aligned}$$

and α_x, α_y and α_{xy} are the transformed thermal coefficients of expansion

$$\begin{aligned}
\alpha_{xx} &= \alpha_1\cos^2\theta + \alpha_2\sin^2\theta \\
\alpha_{yy} &= \alpha_1\sin^2\theta + \alpha_2\cos^2\theta \\
\alpha_{xy} &= 2(\alpha_1 - \alpha_2)\sin\theta\ \cos\theta \qquad (3.4-7)
\end{aligned}$$

The stress–strain relations (5) form the basis for the computation of laminate stiffnesses in the global coordinates (x, y, z).

Example 3.4–1

The material properties of graphite–cpoxy (AS) layers are (see Table 3.4–1):

$$E_1 = 20\times10^6\text{psi}\ ,\ E_2 = 1.3\times10^6\text{psi}\ ,\ \ G_{12} = 1.03\times10^6\text{psi}$$

$$G_{13} = 1.03\times10^6\text{psi}\ ,\ G_{23} = 0.90\times10^6\text{psi}\ ,\ \nu_{12} - 0.3 \quad (3.4-8)$$

The matrix of plane stress–reduced elastic coefficients for the material can be calculated using Eq. (3b) as

$$[Q] = \begin{bmatrix} 20.118 & 0.392 & 0 \\ 0.392 & 1.307 & 0 \\ 0 & 0 & 1.03 \end{bmatrix} \text{ msi} \qquad (3.4-9)$$

The transformed coefficients for various angles of orientation are given below:

$$[\bar{Q}]_{90} = \begin{bmatrix} 1.307 & 0.392 & 0 \\ 0.392 & 20.118 & 0 \\ 0 & 0 & 1.03 \end{bmatrix} \text{ msi} \qquad (3.4-10)$$

$$[\bar{Q}]_{\pm 45} = \begin{bmatrix} 6.582 & 4.522 & \pm 4.702 \\ 4.522 & 6.582 & \pm 4.702 \\ \pm 4.702 & \pm 4.702 & 5.160 \end{bmatrix} \text{ msi} \qquad (3.4-11)$$

$$[\bar{Q}]_{\pm 30} = \begin{bmatrix} 12.318 & 3.490 & \pm 5.861 \\ 3.490 & 2.912 & \pm 2.284 \\ \pm 5.861 & \pm 2.284 & 4.127 \end{bmatrix} \text{ msi} \qquad (3.4-12)$$

Table 3.4–1 Values of the engineering constants for several materials* (the properties of polymeric materials vary with processing and therefore the values given here may differ from those found elsewhere).

Material	E_1	E_2	G_{12}	G_{13}	G_{23}	ν_{12}
Aluminum	10.4	10.4	03.38	03.38	03.38	0.33
Copper	17.0	17.0	06.39	06.39	06.39	0.33
Steel	29.0	29.0	11.24	11.24	11.24	0.29
Gr.–Ep (AS)	20.0	01.3	01.03	01.03	00.90	0.30
Gr.–Ep (T)	19.0	01.5	01.00	00.90	00.90	0.22
Gl.–Ep	07.8	02.6	01.25	01.25	00.50	0.25
Br.–Ep	30.0	03.0	01.50	01.50	00.60	0.25

* Values of the moduli are in msi=million psi; 1 psi = 6.895 kN/m^2.

Gr.–Ep (AS) = graphite–epoxy (AS/3501) ; Gr.–Ep (T) = graphite–epoxy (T300/934) ; Gl.–Ep = glass–epoxy ; Br.–Ep = boron–epoxy.

EXERCISE PROBLEMS

3.1 Express the stress–strain relations in Eq. (3.2–4) in terms of the engineering constants [*i.e.*, find the inverse of Eq. (3.2–11)].

3.2 Express the stress–strain relations in Eq. (3.2–8) in terms of the engineering constants (*Hint:* Use the result of Problem 3.1).

3.3 Verify Eq. (3.3–8a) by carrying out the matrix operation in Eq. (3.3–4) and rewriting the result in single–column format for stresses and strains.

3.4 Verify Eq. (3.3–10) by carrying out the matrix operation,

$$[c] = [A]^T[\bar{c}][A]$$

and rewriting the result in single–column format for strains and stresses.

3.5 Establish the relations in Eq. (3.4–2) and verify the result in Eq. (3.4–3a,b).

3.6 Compute the transformed stiffnesses $\bar{Q}_{ij}$ for $\theta = 22.5°$ for the graphite–epoxy (AS) material layers.

REFERENCES FOR ADDITIONAL READING

1. Agarwal, B. D. and Broutman, L. J., *Analysis and Performance of Fiber Composites*, John Wiley & Sons, New York, 1980.

2. Ambartsumian, S. A., *Theory of Anisotropic Shells*, NASA Report TT F–118, 1964.

3. Ambartsumian, S. A., *Theory of Anisotropic Plates*, Izdat. Nauka, Moskva (1967), English translation by Technomic, Stamford, CT, 1969.

4. Christensen, R. M., *Mechanics of Composite Materials*, John Wiley & Sons, New York, 1979.

5. Lekhnitskii, S. G., *Theory of Elasticity of an Anisotropic Body*, Mir Publishers, Moscow (Translated from the revised 1977 Russian Edition), 1982.

6. Jones, R. M., *Mechanics of Composite Materials*, Scripta Book Co., Washington, D. C., 1975.

7. Mallick, P. K., *Fiber–Reinforced Composites. Materials, Manufacturing, and Design*, Marcel Dekker, New York, 1988.

8. Reddy, J. N., *Energy and Variational Methods in Applied Mechanics*, John Wiley & Sons, New York, 1984.

9. Reddy, J. N., *Mechanics of Laminated Composite Structures: Theory and Analysis,* Lecture Notes, Virginia Polytechnic Institute and State University, Blacksburg, VA 24061, October 1988; also see Chapters 14 and 15 in: *Finite Element Analysis for Engineering Design*, J. N. Reddy, C. S. Krishna Moorthy, and K. N. Seetharamu (eds.), Vol. 37, Springer–Verlag, Berlin, 1988.

10. Tsai, S. W. and Hahn, H. T., *Introduction to Composite Materials*, Technomic, Westport, CT, 1980.

11. Vasiliev, V. V., *Mechanics of Composite Structures*, Taylor & Francis, Washington, D. C., 1993.

12. Whitney, J. M., *Structural Analysis of Laminated Anisotropic Plates*, Technomic, Lancaster, PA, 1987.

13. Reddy, J. N. (ed.), *Mechanics of Composite Materials. Selected Works of Nicholas J. Pagano*, Kluwer Academic Publishers, The Netherlands, 1994.

Chapter Four

THEORIES OF LAMINATED PLATES

4.1 INTRODUCTION

A composite lamina is a building block for laminates. Laminae are bonded together to form a laminate with the desired thickness and stiffness. In most applications, the thickness of a laminate is small compared to the planar dimensions (*i.e.*, the side–to–thickness ratio of a laminate is greater than 20). Therefore, laminate theories based on equivalent two–dimensional descriptions are used to analyze laminated composite structures. The two–dimensional theories, termed *equivalent–single layer theories*, are obtained from the three–dimensional elasticity theory by making assumptions concerning the variation of displacements and/or stresses through the thickness of the laminate.

In the *classical laminated plate theory* (CLPT), which is an extension of the classical plate theory to laminated plates, the inplane displacements are assumed to vary linearly through the thickness and the transverse displacement is assumed to be constant through the thickness (*i.e.*, transverse normal strain is zero). The classical laminate theory is found to be adequate for most applications where the thickness of the laminate is small, by two orders of magnitude compared to the inplane dimensions. When the classical laminate theory is not applicable, a refined theory that accounts for the transverse shear strains is used. The *first–order shear deformation theory* (FSDT) accounts for constant state of transverse shear strains through the thickness. There are higher–order, equivalent–single–layer shear deformation theories [1–5] as well as layerwise theories [6,7], which are not covered here.

The objective of this chapter is to present the governing equations of the classical laminate theory and the first–order shear deformation theory. In presenting the equations, we give the basic assumptions, the assumed displacement field, associated strains and stresses, and governing differential equations of each theory. The derivation of the equations of motion is based on the *principle of virtual displacements* (see Reddy [1]). However, the details of the actual derivation will be omitted, and the results will be summarized.

4.2 CLASSICAL LAMINATED PLATE THEORY

4.2–1 Introduction

The classical laminated plate theory is an extension of the classical plate theory of plates to laminated composite plates. In the classical plate theory one assumes that straight lines perpendicular to the midplane before deformation remain (1) straight, (2) inextensible, and (3) normal to the midplane after deformation. These three assumptions, collectively known as the *Kirchhoff–Love hypothesis,* lead to zero transverse shear strains (ε_{xz}, ε_{yz}) and transverse normal strain (ε_{zz}). Since the (virtual) work done by the actual internal forces of the plate in moving through the virtual displacements is the product of forces associated with the stress field and the (virtual) displacements, which result in (virtual) strains, the (virtual) work done by the transverse stresses is zero because of the zero transverse (virtual) strains. Consequently, transverse stresses σ_{xz}, σ_{yz} and σ_{zz} do not contribute to the equations of motion. This amounts to omitting the transverse stresses in the classical laminate theory. This is equivalent to, from a physical point of view, assuming that the plate is "infinitely" rigid in the transverse direction. When plates are very thin, this assumption is not so bad as it may appear on the surface.

4.2–2 Displacements and Strains

Consider a laminate of total thickness h composed of N orthotropic layers. A typical lamina, say the kth layer ($k = 1, 2, \cdots, L$), has a uniform thickness h_k, material properties E_1^k, E_2^k, etc., and its principal material coordinates oriented at an angle $\theta_1 = k$ to the laminate (global) coordinate, x. In deriving the governing equations, we make the following assumptions:

(i) the layers are perfectly bonded together,

(ii) the material of each layer is linearly elastic and has three planes of material symmetry (*i.e.*, orthotropic),

(iii) each layer is of uniform thickness,

(iv) the strains are small, and

(v) the Kirchhoff–Love hypothesis holds.

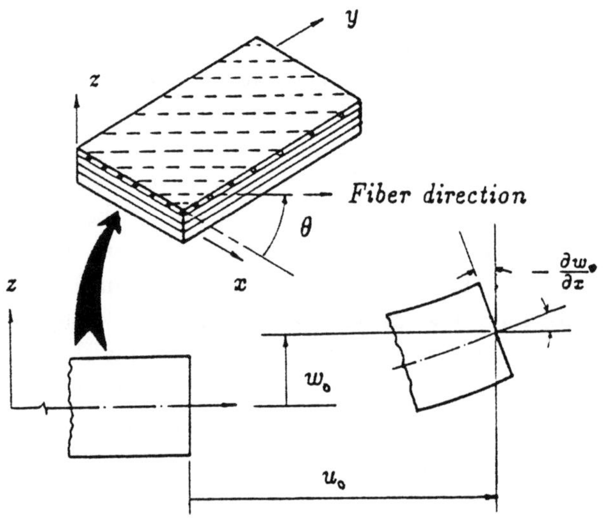

Figure 4.2–1: Bending of laminated plates under the Love–Kirchhoff hypothesis.

The laminate coordinate system is chosen such that the xy–plane coincides with the midplane of the laminate. Although this is not necessary, it makes the derivations simple and consistent with the approach taken in a vast majority of the literature. The assumptions of the Kirchhoff hypothesis require that the displacements (u, v, w) be such that (see Figure 4.2–1)

$$
\begin{aligned}
u(x,y,z,t) &= u_0(x,y,t) - z\frac{\partial w_0}{\partial x}(x,y,t) \\
v(x,y,z,t) &= v_0(x,y,t) - z\frac{\partial w_0}{\partial y}(x,y,t) \\
w(x,y,z,t) &= w_0(x,y,t)
\end{aligned}
\tag{4.2 - 1}
$$

where (u_0, v_0, w_0) are the displacements of a point on the xy–plane (or midplane), and t denotes time.

It is important to note that, in modelling a laminated plate composed of multiple layers of possibly dissimilar–material layers, we tacitly assumed that the strains are continuous through the thickness, including the interfaces of dissimilar–material layers. This assumption plays a significant role in developing laminate theories; it allows us to replace a laminate with an equivalent single layer whose material coefficients are averaged over the laminate thickness, as will be shown in the following sections.

The strains in the laminate can be computed using the strain–displacement relations [see Eq. (2.5–1)]

$$
\begin{aligned}
\varepsilon_{xx} \equiv \varepsilon_{11} &= \frac{\partial u}{\partial x} \\
\varepsilon_{yy} \equiv \varepsilon_{22} &= \frac{\partial v}{\partial y} \\
\varepsilon_{zz} \equiv \varepsilon_{33} &= \frac{\partial w}{\partial z} \\
\gamma_{xy} \equiv 2\varepsilon_{12} &= \frac{\partial u}{\partial y} + \frac{\partial v}{\partial x} \\
\gamma_{xz} \equiv 2\varepsilon_{13} &= \frac{\partial u}{\partial z} + \frac{\partial w}{\partial x} \\
\gamma_{yz} \equiv 2\varepsilon_{23} &= \frac{\partial v}{\partial z} + \frac{\partial w}{\partial y}
\end{aligned}
\tag{4.2 - 2}
$$

Using the displacement field (1) in (2), we obtain

$$\varepsilon_{xx} = \frac{\partial u_0}{\partial x} - z\frac{\partial^2 w_0}{\partial x^2}$$
$$\varepsilon_{yy} = \frac{\partial v_0}{\partial y} - z\frac{\partial^2 w_0}{\partial y^2}$$
$$\gamma_{xy} = \frac{\partial u_0}{\partial y} + \frac{\partial v_0}{\partial x} - 2z\frac{\partial^2 w_0}{\partial x \partial y}$$
$$\gamma_{yz} = 0, \ \gamma_{xz} = 0, \ \varepsilon_{zz} = 0 \qquad (4.2-3)$$

In matrix notation, Eq. (3) takes the form

$$\left\{ \begin{array}{c} \varepsilon_{xx} \\ \varepsilon_{yy} \\ \varepsilon_{xy} \end{array} \right\} = \left\{ \begin{array}{c} \frac{\partial u_0}{\partial x} \\ \frac{\partial v_0}{\partial y} \\ \frac{\partial u_0}{\partial y} + \frac{\partial v_0}{\partial x} \end{array} \right\} - z \left\{ \begin{array}{c} \frac{\partial^2 w_0}{\partial x^2} \\ \frac{\partial^2 w_0}{\partial y^2} \\ 2\frac{\partial^2 w_0}{\partial x \partial y} \end{array} \right\} \qquad (4.2-4)$$

We note from Eq. (4) that the strains vary linearly through the laminate thickness, and they are continuous at the interface of layers, as noted earlier. For a fixed value of z, the strains vary only with respect to the x and y coordinates.

4.2–3 Equations of Motion

The equations of motion can be derived using the principle of virtual displacements for the dynamic case [1]. The principle states that the virtual work done by the actual forces (internal as well as external) in moving through an admissible virtual displacements (*i.e.*, those that satisfy the geometric constraints of the problem) is zero if the body is in static or dynamic equilibrium. The application of the principle to the case at hand yields the following equations of motion:

$$\frac{\partial N_{xx}}{\partial x} + \frac{\partial N_{xy}}{\partial y} = I_0 \ddot{u}_0 - I_1 \frac{\partial \ddot{w}_0}{\partial x} \qquad (4.2-4)$$

$$\frac{\partial N_{xy}}{\partial x} + \frac{\partial N_{yy}}{\partial y} = I_0 \ddot{v}_0 - I_1 \frac{\partial \ddot{w}_0}{\partial y} \qquad (4.2-5)$$

$$\frac{\partial^2 M_{xx}}{\partial x^2}+2\frac{\partial^2 M_{xy}}{\partial y \partial x}+\frac{\partial^2 M_{yy}}{\partial y^2}+q+\hat{N}_{xx}\frac{\partial^2 w_0}{\partial x^2}+2\hat{N}_{xy}\frac{\partial^2 w_0}{\partial x \partial y}+\hat{N}_{yy}\frac{\partial^2 w_0}{\partial y^2}$$
$$= I_0\ddot{w}_0 + I_2\left(\frac{\partial^2 \ddot{w}_0}{\partial x^2}+\frac{\partial^2 \ddot{w}_0}{\partial y^2}\right) - I_1\left(\frac{\partial \ddot{u}_0}{\partial x}+\frac{\partial \ddot{v}_0}{\partial y}\right) \quad (4.2-6)$$

where, for example, $\ddot{w}_0 \equiv \partial^2 w_0/\partial t^2$, q is the distributed transverse load, $\hat{N}_{xx}, \hat{N}_{yy}$, and $\hat{N}_{xy}$ are inplane applied forces, and (N's, M's) are the force and moment resultants (see Figure 4.2–2)

$$\begin{Bmatrix} N_{xx} \\ N_{yy} \\ N_{xy} \end{Bmatrix} = \int_{-\frac{h}{2}}^{\frac{h}{2}} \begin{Bmatrix} \sigma_{xx} \\ \sigma_{yy} \\ \sigma_{xy} \end{Bmatrix} dz \, , \quad \begin{Bmatrix} M_{xx} \\ M_{yy} \\ M_{xy} \end{Bmatrix} = \int_{-\frac{h}{2}}^{\frac{h}{2}} \begin{Bmatrix} \sigma_{xx} \\ \sigma_{yy} \\ \sigma_{xy} \end{Bmatrix} z \, dz \quad (4.2-7)$$

and I_i are the mass inertias

$$\begin{Bmatrix} I_0 \\ I_1 \\ I_2 \end{Bmatrix} = \int_{-\frac{h}{2}}^{\frac{h}{2}} \begin{Bmatrix} 1 \\ z \\ z^2 \end{Bmatrix} \rho \, dz \quad (4.2-8)$$

The second term on the right side of the equality in Eqs. (4) and (5) is called the *rotatory inertia* term, and it is often omitted because it is small in comparison to the $I_0\ddot{u}$ and $I_0\ddot{v}$ terms. The same comment applies to the second term on the right hand side of Eq. (6).

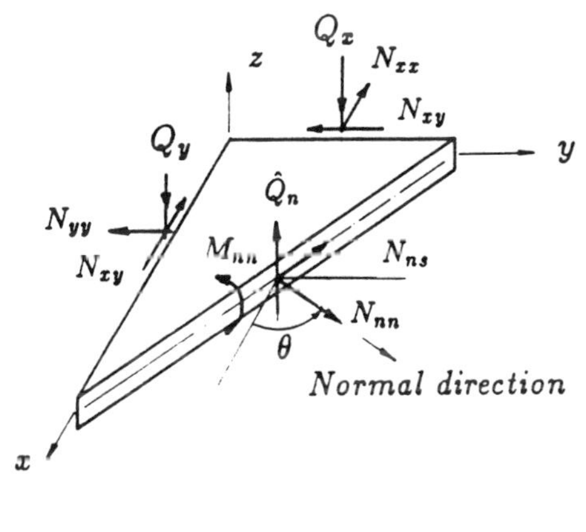

Figure 4.2–2: Stress resultants on an arbitrary edge of a plate.

The geometric and force boundary conditions of the theory are given by (see Reddy [1]):

Geometric Boundary Conditions: Specify

$$u_{nn}, \quad u_{ns}, \quad w_0, \quad \frac{\partial w_0}{\partial n} \tag{4.2 – 9}$$

Force Boundary Conditions: Specify

$$N_{nn}, \quad N_{ns}, \quad \hat{Q}_n, \quad M_{nn} \tag{4.2 – 10}$$

where u_{nn} is the inplane displacement normal to the boundary, u_{ns} is the inplane displacement tangential to the boundary, and (N_{nn}, N_{ns}) are the corresponding edge forces (per unit thickness); Q_n and M_n are the transverse edge force and edge bending moment about the tangent to the edge, respectively (see Figure 4.2–2):

$$u_{nn} = u_0 n_x + v_0 n_y, \quad u_{ns} = -u_0 n_y + v_0 n_x$$

$$\frac{\partial w_0}{\partial n} = \frac{\partial w_0}{\partial x} n_x + \frac{\partial w_0}{\partial y} n_y \tag{4.2 – 11}$$

$$N_{nn} = n_x^2 N_{xx} + n_y^2 N_{yy} + 2n_x n_y N_{xy} \tag{4.2 – 12a}$$

$$N_{ns} = (N_{yy} - N_{xx}) n_x n_y + N_{xy}(n_x^2 - n_y^2) \tag{4.2 – 12b}$$

$$\hat{Q}_n = n_x Q_x + n_y Q_y - \frac{\partial M_{ns}}{\partial s} \tag{4.2 – 13}$$

and

$$\begin{Bmatrix} Q_x \\ Q_y \end{Bmatrix} = \int_{-\frac{h}{2}}^{\frac{h}{2}} \begin{Bmatrix} \sigma_{xz} \\ \sigma_{yz} \end{Bmatrix} dz \tag{4.2 – 14}$$

$$M_{nn} = n_x^2 M_{xx} + n_y^2 M_{yy} + 2n_x n_y M_{xy}$$
$$M_{ns} = (M_{yy} - M_{xx}) n_x n_y + M_{xy}(n_x^2 - n_y^2) \tag{4.2 – 15}$$

and n_x and n_y are the direction cosines of the unit normal on the boundary of the plate, and where Q_x and Q_y are the transverse shear forces (per unit thickness) on the edge perpendicular to x and y axes, respectively.

The quantities appearing in the geometric boundary conditions are termed *primary variables,* and those appearing in the force boundary conditions are known as the *secondary variables.* Thus, the generalized displacements u_0, v_0 (or u_{nn} and u_{ns}), w_0, and $\frac{\partial w_0}{\partial n}$ are the primary variables, and N_{nn}, N_{ns}, $\hat{Q}_n$, and M_{nn} are the secondary variables of the classical plate theory.

4.2–4 Laminate Constitutive Equations

In view of the fact that $(\sigma_{xz}, \sigma_{yz}, \sigma_{zz})$ are neglected in the classical laminate theory and that the thickness is small compared to the inplane dimensions of the laminate, it is assumed that the laminate is in a state of plane stress (see Section 3.4). Hence, the stresses and strains in a typical orthotropic layer are related according to [see Eq. (3.4–5)],

$$\begin{Bmatrix} \sigma_{xx} \\ \sigma_{yy} \\ \sigma_{xy} \end{Bmatrix}^k = \begin{bmatrix} \bar{Q}_{11} & \bar{Q}_{12} & \bar{Q}_{16} \\ \bar{Q}_{12} & \bar{Q}_{22} & \bar{Q}_{26} \\ \bar{Q}_{16} & \bar{Q}_{26} & \bar{Q}_{66} \end{bmatrix}^k \begin{Bmatrix} \varepsilon_{xx} - \alpha_{xx}\Delta T \\ \varepsilon_{yy} - \alpha_{yy}\Delta T \\ 2\varepsilon_{xy} - \alpha_{xy}\Delta T \end{Bmatrix} \qquad (4.2-16)$$

where $\bar{Q}_{ij}$ are the transformed, plane–stress–reduced stiffnesses [see Eqs. (3.4–3)–(3.4–6)], and α_{xx}, α_{yy} and α_{xy} are the transformed thermal coefficients of expansion (see Eq. (3.4–7)]. The laminate constitutive equations relate the force and moment resultants (N's,M's) to the displacement gradients of Eq. (4.2–5). The definitions of the force and moment resultants in Eq. (4.2–8) involve integration over the total thickness of the laminate. However, from Eq. (16) it is clear that the stresses are discontinuous (because Q's change from layer to layer while the strains are continuous), requiring the integration to be carried out layer by layer. We have

$$\begin{Bmatrix} N_{xx} \\ N_{yy} \\ N_{xy} \end{Bmatrix} = \begin{bmatrix} A_{11} & A_{12} & A_{16} \\ A_{12} & A_{22} & A_{26} \\ A_{16} & A_{26} & A_{66} \end{bmatrix} \begin{Bmatrix} \frac{\partial u_0}{\partial x} \\ \frac{\partial v_0}{\partial y} \\ \frac{\partial u_0}{\partial y} + \frac{\partial v_0}{\partial x} \end{Bmatrix}$$

$$
- \begin{bmatrix} B_{11} & B_{12} & B_{16} \\ B_{12} & B_{22} & B_{26} \\ B_{16} & B_{26} & B_{66} \end{bmatrix} \begin{Bmatrix} \frac{\partial^2 w_0}{\partial x^2} \\ \frac{\partial^2 w_0}{\partial y^2} \\ 2\frac{\partial^2 w_0}{\partial x \partial y} \end{Bmatrix} - \begin{Bmatrix} N_{xx}^T \\ N_{yy}^T \\ N_{xy}^T \end{Bmatrix} \quad (4.2-17)
$$

$$
\begin{Bmatrix} M_{xx} \\ M_{yy} \\ M_{xy} \end{Bmatrix} = \begin{bmatrix} B_{11} & B_{12} & B_{16} \\ B_{12} & B_{22} & B_{26} \\ B_{16} & B_{26} & B_{66} \end{bmatrix} \begin{Bmatrix} \frac{\partial u_0}{\partial x} \\ \frac{\partial v_0}{\partial y} \\ \frac{\partial u_0}{\partial y} + \frac{\partial v_0}{\partial x} \end{Bmatrix}
$$

$$
- \begin{bmatrix} D_{11} & D_{12} & D_{16} \\ D_{12} & D_{22} & D_{26} \\ D_{16} & D_{26} & D_{66} \end{bmatrix} \begin{Bmatrix} \frac{\partial^2 w_0}{\partial x^2} \\ \frac{\partial^2 w_0}{\partial y^2} \\ 2\frac{\partial^2 w_0}{\partial x \partial y} \end{Bmatrix} - \begin{Bmatrix} M_{xx}^T \\ M_{yy}^T \\ M_{xy}^T \end{Bmatrix} \quad (4.2-18)
$$

where A_{ij} denote the extensional stiffnesses, D_{ij} the bending stiffnesses, and B_{ij} the bending–extensional coupling stiffnesses of a laminate

$$
(A_{ij}, B_{ij}, D_{ij}) = \sum_{k=1}^{L} \int_{z_k}^{z_{k+1}} \bar{Q}_{ij}^{(k)} (1, z, z^2) dz \qquad (4.2-19a)
$$

and $\{N^T\}$ and $\{M^T\}$ are thermal stress resultants

$$
\{N^T\} = \sum_{k=1}^{L} \int_{z_k}^{z_{k+1}} [\bar{Q}]^{(k)} \{\alpha\}^{(k)} \Delta T \, dz
$$

$$
\{M^T\} = \sum_{k=1}^{L} \int_{z_k}^{z_{k+1}} [\bar{Q}]^{(k)} \{\alpha\}^{(k)} \Delta T \, z \, dz \qquad (4.2-19b)
$$

Here (z_k, z_{k+1}) denote the thickness coordinates of the bottom and top, respectively, of the kth layer, $\bar{Q}_{ij}^{(k)}$ are the material stiffnesses of the kth layer, referred to the laminate coordinates [see Eq. (3.4–6) and (3.4–3b)], and L denotes the total number of layers.

Throughout this book it will be assumed that the material stiffnesses $Q_{ij}^{(k)}$ of individual plies are constant (*i.e.*, independent of position) and orthotropic with respect to the principal material coordinates. Then $\bar{Q}_{ij}^{(k)}$ are constant, and possibly anisotropic within each layer. Hence, the integration in Eq. (19a) over the thickness can be carried out easily. We obtain

$$A_{ij} = \sum_{k=1}^{L} \bar{Q}_{ij}^{(k)} (z_{k+1} - z_k) \tag{4.2 – 20a}$$

$$B_{ij} = \frac{1}{2} \sum_{k=1}^{L} \bar{Q}_{ij}^{(k)} (z_{k+1}^2 - z_k^2) \tag{4.2 – 20b}$$

$$D_{ij} = \frac{1}{3} \sum_{k=1}^{L} \bar{Q}_{ij}^{(k)} (z_{k+1}^3 - z_k^3) \tag{4.2 – 20c}$$

4.2–5 Equations of Motion in Terms of Displacements

The equations of motion in Eqs. (5)–(7) can be expressed in terms of the displacements (u_0, v_0, w_0) for the purpose of developing either analytical or displacement finite element models (see Chapter 5). The force and moment resultants in terms of the displacements from Eqs. (17) and (18) are introduced into the equations of motion (5)–(7) to obtain a system of three differential equations in terms of three displacements (u_0, v_0, w_0). Note that these equations are second order in u_0 and v_0 and fourth order in w_0. We have

$$\begin{aligned}
&\frac{\partial}{\partial x}\left[A_{11}\frac{\partial u_0}{\partial x} + A_{12}\frac{\partial u_0}{\partial y} + A_{16}\left(\frac{\partial u_0}{\partial y} + \frac{\partial v_0}{\partial x}\right) - B_{11}\frac{\partial^2 w_0}{\partial x^2}\right.\\
&\left. -B_{12}\frac{\partial^2 w_0}{\partial y^2} - 2B_{16}\frac{\partial^2 w_0}{\partial x \partial y}\right] + \frac{\partial}{\partial y}\left[A_{16}\frac{\partial u_0}{\partial x} + A_{26}\frac{\partial u_0}{\partial y}\right.\\
&\left. +A_{66}\left(\frac{\partial u_0}{\partial y} + \frac{\partial v_0}{\partial x}\right) - B_{16}\frac{\partial^2 w_0}{\partial x^2} - B_{26}\frac{\partial^2 w_0}{\partial y^2} - 2B_{66}\frac{\partial^2 w_0}{\partial x \partial y}\right]\\
&- \frac{\partial N_{xx}^T}{\partial x} - \frac{\partial N_{xy}^T}{\partial y} = I_0\frac{\partial^2 u_0}{\partial t^2} - I_1\frac{\partial^3 w_0}{\partial x \partial t^2}
\end{aligned} \tag{4.2 – 21}$$

$$\frac{\partial}{\partial x}\left[A_{16}\frac{\partial u_0}{\partial x}+A_{26}\frac{\partial v_0}{\partial y}+A_{66}\left(\frac{\partial u_0}{\partial y}+\frac{\partial v_0}{\partial x}\right)-B_{16}\frac{\partial^2 w_0}{\partial x^2}\right.$$
$$\left.-B_{26}\frac{\partial^2 w_0}{\partial y^2}-2B_{66}\frac{\partial^2 w_0}{\partial x\partial y}\right]+\frac{\partial}{\partial y}\left[A_{12}\frac{\partial u_0}{\partial x}+A_{22}\frac{\partial v_0}{\partial y}\right.$$
$$\left.+A_{26}\left(\frac{\partial u_0}{\partial y}+\frac{\partial v_0}{\partial x}\right)-B_{12}\frac{\partial^2 w_0}{\partial x^2}-B_{22}\frac{\partial^2 w_0}{\partial y^2}-2B_{26}\frac{\partial^2 w_0}{\partial x\partial y}\right]$$
$$-\frac{\partial N_{xy}^T}{\partial x}-\frac{\partial N_{yy}^T}{\partial y}=I_0\frac{\partial^2 v_0}{\partial t^2}-I_1\frac{\partial^3 w_0}{\partial y\partial t^2} \qquad (4.2-22)$$

$$\frac{\partial^2}{\partial x^2}\left[B_{11}\frac{\partial u_0}{\partial x}+B_{12}\frac{\partial v_0}{\partial y}+B_{16}\left(\frac{\partial u_0}{\partial y}+\frac{\partial v_0}{\partial x}\right)-D_{11}\frac{\partial^2 w_0}{\partial x^2}\right.$$
$$\left.-D_{12}\frac{\partial^2 w_0}{\partial y^2}-2D_{16}\frac{\partial^2 w_0}{\partial x\partial y}\right]+2\frac{\partial^2}{\partial x\partial y}\left[B_{16}\frac{\partial u_0}{\partial x}+B_{26}\frac{\partial v_0}{\partial y}+\right.$$
$$\left.B_{66}\left(\frac{\partial u_0}{\partial y}+\frac{\partial v_0}{\partial x}\right)-D_{16}\frac{\partial^2 w_0}{\partial x^2}-D_{26}\frac{\partial^2 w_0}{\partial y^2}-2D_{66}\frac{\partial^2 w_0}{\partial x\partial y}\right]$$
$$+\frac{\partial^2}{\partial y^2}\left[B_{12}\frac{\partial u_0}{\partial x}+B_{22}\frac{\partial v_0}{\partial y}+B_{26}\left(\frac{\partial u_0}{\partial y}+\frac{\partial v_0}{\partial x}\right)\right.$$
$$\left.-D_{12}\frac{\partial^2 w_0}{\partial x^2}-D_{22}\frac{\partial^2 w_0}{\partial y^2}-2D_{26}\frac{\partial^2 w_0}{\partial x\partial y}\right]+q$$
$$+\hat{N}_{xx}\frac{\partial^2 w_0}{\partial x^2}+2\hat{N}_{xy}\frac{\partial^2 w_0}{\partial x\partial y}+\hat{N}_{yy}\frac{\partial^2 w_0}{\partial y^2}$$
$$-\frac{\partial^2 M_{xx}^T}{\partial x^2}-2\frac{\partial^2 M_{xy}^T}{\partial y\partial x}-\frac{\partial^2 M_{yy}^T}{\partial y^2}=I_0\frac{\partial^2 w_0}{\partial t^2}$$
$$-I_2\frac{\partial^2}{\partial t^2}\left(\frac{\partial^2 w_0}{\partial x^2}+\frac{\partial^2 w_0}{\partial y^2}\right)+I_1\frac{\partial^2}{\partial t^2}\left(\frac{\partial u_0}{\partial x}+\frac{\partial v_0}{\partial y}\right) \qquad (4.2-23)$$

4.3 THE FIRST–ORDER SHEAR DEFORMATION THEORY

4.3–1 Displacement Field and Strains

The classical plate theory is adequate for the analysis of thin laminates, especially when the transverse deformation is negligible. However, laminates made of advanced fiber–reinforced composite materials, whose elastic to shear modulus ratios (E_1/G_{13} and E_1/G_{23}) are very large, are susceptible to thickness failures because their effective transverse shear moduli (G_{13} and G_{23}) are significantly smaller than the effective elastic moduli (E_1) along the fiber direction. Although the transverse shear and normal stresses ($\sigma_{xz}, \sigma_{yz}, \sigma_z$) are an order of magnitude smaller than the inplane stresses ($\sigma_x, \sigma_y, \sigma_{xy}$), the material strength *allowables* for the transverse stresses are also an order of magnitude smaller than the allowables for the inplane stresses. Thus, the classical laminate theory developed here is not to be used for composites that are likely to fail in transverse shear or delamination.

The most widely used displacement based shear deformation theory is the first–order shear deformation theory. In this theory the normality condition of the Kirchhoff–Love hypothesis is removed, allowing for independent rotation of transverse normals. Thus, straight line normal to the midplane of the laminate remains straight and inextensible after deformation, but it does not necessarily remain perpendicular to the midsurface. Thus, the displacement field of the first–order shear deformation theory is assumed to be of the form

$$\begin{aligned} u_1(x,y,z) &= u_0(x,y) + z\phi_x(x,y) \\ u_2(x,y,z) &= v_0(x,y) + z\phi_y(x,y) \\ u_3(x,y,z) &= w_0(x,y) \end{aligned} \qquad (4.3-1)$$

where (ϕ_x, ϕ_y) are the rotations of the transverse normal about the (y, x)–axes, respectively.

The strains associated with the displacement field (1) are

$$\begin{aligned} \varepsilon_{xx} &= \frac{\partial u_0}{\partial x} + z\frac{\partial \phi_x}{\partial x} \\ \varepsilon_{yy} &= \frac{\partial v_0}{\partial y} + z\frac{\partial \phi_y}{\partial y} \\ \gamma_{xy} &= \frac{\partial u_0}{\partial y} + \frac{\partial v_0}{\partial x} + z\left(\frac{\partial \phi_x}{\partial y} + \frac{\partial \phi_y}{\partial x}\right) \end{aligned}$$

$$\gamma_{xz} = \phi_x + \frac{\partial w_0}{\partial x}, \quad \gamma_{yz} = \phi_y + \frac{\partial w_0}{\partial y}, \quad \varepsilon_{zz} = 0 \qquad (4.3-2)$$

4.3–2 Equations of Motion

The equations of motion of the first–order shear deformation theory are

$$\frac{\partial N_{xx}}{\partial x} + \frac{\partial N_{xy}}{\partial y} = I_0 \frac{\partial^2 u_0}{\partial t^2} + I_1 \frac{\partial^2 \phi_x}{\partial t^2} \qquad (4.3-3)$$

$$\frac{\partial N_{xy}}{\partial x} + \frac{\partial N_{yy}}{\partial y} = I_0 \frac{\partial^2 v_0}{\partial t^2} + I_1 \frac{\partial^2 \phi_y}{\partial t^2} \qquad (4.3-4)$$

$$\frac{\partial Q_x}{\partial x} + \frac{\partial Q_y}{\partial y} + q + \hat{N}_{xx} \frac{\partial^2 w_0}{\partial x^2} + 2\hat{N}_{xy} \frac{\partial^2 w_0}{\partial x \partial y} + \hat{N}_{yy} \frac{\partial^2 w_0}{\partial y^2} = I_0 \frac{\partial^2 w_0}{\partial t^2} \qquad (4.3-5)$$

$$\frac{\partial M_{xx}}{\partial x} + \frac{\partial M_{xy}}{\partial y} - Q_x = I_1 \frac{\partial^2 u_0}{\partial t^2} + I_0 \frac{\partial^2 \phi_x}{\partial t^2} \qquad (4.3-6)$$

$$\frac{\partial M_{xy}}{\partial x} + \frac{\partial M_{yy}}{\partial y} - Q_y = I_1 \frac{\partial^2 v_0}{\partial t^2} + I_0 \frac{\partial^2 \phi_y}{\partial t^2} \qquad (4.3-7)$$

where (N's,M's,Q's) are the force and moment resultants defined in Eqs. (4.2-8) and (4.2-14).

The geometric and force boundary conditions of the first–order shear deformation theory are given by (see Reddy [1]):

Geometric Boundary Conditions: Specify

$$u_{nn}, \quad u_{ns}, \quad w_0, \quad \phi_{nn}, \quad \phi_{ns} \qquad (4.3-8)$$

Force Boundary Conditions: Specify

$$N_{nn}, \quad N_{ns}, \quad Q_n, \quad M_{nn}, \quad M_{ns} \qquad (4.3-9)$$

where all quantities were defined in Eqs. (4.2–12)–(4.2–15), except

$$Q_n = n_x Q_x + n_y Q_y \tag{4.3 – 10a}$$

$$\phi_{nn} = \phi_x n_x + \phi_y n_y, \quad \phi_{ns} = -\phi_x n_y + \phi_y n_x \tag{4.3 – 10b}$$

Thus, the primary variables of the first–order theory are u_{nn}, u_{ns}, w_0, ϕ_{nn}, and ϕ_{ns}, and the secondary variables are N_{nn}, N_{ns}, Q_n, M_{nn}, and M_{ns}.

The first–order shear deformation theory is also referred to as the "Mindlin plate theory" [8], although the original idea of such an expansion can be found in earlier works by Basset [9], Hildebrand, Reissner, and Thomas [10], and Hencky [11]. The first–order shear deformation theory yields a constant value of transverse shear strain through the thickness of the plate and thus requires shear correction factors. The shear correction factors are dimensionless quantities introduced to account for the discrepancy between the constant state of shear strains in the first–order shear deformation theory and the actual distribution of shear strains in the elasticity theory. For composite laminates, the shear correction factors, in general, depend on the constituent ply properties, lamination scheme, and type of structure (*i.e.*, geometry and boundary conditions) [12–15].

The transverse shear forces (Q_x, Q_y) can be expressed in terms of the displacements as

$$\left\{ \begin{matrix} Q_x \\ Q_y \end{matrix} \right\} = \int_{-\frac{h}{2}}^{\frac{h}{2}} \left\{ \begin{matrix} \sigma_{xz} \\ \sigma_{yz} \end{matrix} \right\} dz = K \begin{bmatrix} A_{55} & A_{45} \\ A_{45} & A_{44} \end{bmatrix} \left\{ \begin{matrix} \phi_x + \frac{\partial w_0}{\partial x} \\ \phi_y + \frac{\partial w_0}{\partial y} \end{matrix} \right\} \tag{4.3 – 11}$$

where

$$\begin{aligned} A_{ij} &= \int_{-\frac{h}{2}}^{\frac{h}{2}} Q_{ij}^{(k)} dz \quad (i, j = 4, 5) \\ &= \sum_{k=1}^{N} Q_{ij}^{(k)} (z_{k+1} - z_k) \end{aligned} \tag{4.3 – 12}$$

and K is the shear correction coefficient (see [15]). The material stiffnesses $\bar{Q}_{ij}^{(k)}$ $(i, j = 4, 5)$ in the laminate coordinates are related to the $Q_{ij}^{(k)}$ in the material coordinates by

$$\begin{aligned}\bar{Q}_{44} &= Q_{44}\cos^2\theta + Q_{55}\sin^2\theta \\ \bar{Q}_{45} &= (Q_{55}-Q_{44})\cos\theta\ \sin\theta \\ \bar{Q}_{55} &= Q_{55}\cos^2\theta + Q_{44}\sin^2\theta \end{aligned} \qquad (4.3-13)$$

The equations of motion (3)–(7) can be expressed in terms of the generalized displacements (u_0, v_0, w_0, ϕ_x and ϕ_y) as:

$$\begin{aligned}&\frac{\partial}{\partial x}\left[A_{11}\frac{\partial u_0}{\partial x} + A_{12}\frac{\partial v_0}{\partial y} + A_{16}\left(\frac{\partial u_0}{\partial y} + \frac{\partial v_0}{\partial x}\right) + B_{11}\frac{\partial \phi_x}{\partial x} + B_{12}\frac{\partial \phi_y}{\partial y}\right. \\ &\quad \left. + B_{16}\left(\frac{\partial \phi_x}{\partial y} + \frac{\partial \phi_y}{\partial x}\right)\right] + \frac{\partial}{\partial y}\left[A_{16}\frac{\partial u_0}{\partial x} + A_{26}\frac{\partial v_0}{\partial y}\right. \\ &\quad \left. + A_{66}\left(\frac{\partial u_0}{\partial y} + \frac{\partial v_0}{\partial x}\right) + B_{16}\frac{\partial \phi_x}{\partial x} + B_{26}\frac{\partial \phi_y}{\partial y} + B_{66}\left(\frac{\partial \phi_x}{\partial y} + \frac{\partial \phi_y}{\partial x}\right)\right] \\ &\quad - \left(\frac{\partial N_{xx}^T}{\partial x} + \frac{\partial N_{xy}^T}{\partial y}\right) = I_0\frac{\partial^2 u_0}{\partial t^2} + I_1\frac{\partial^2 \phi_x}{\partial t^2} \end{aligned} \qquad (4.3-14)$$

$$\begin{aligned}&\frac{\partial}{\partial x}\left[A_{16}\frac{\partial u_0}{\partial x} + A_{26}\frac{\partial v_0}{\partial y} + A_{66}\left(\frac{\partial u_0}{\partial y} + \frac{\partial v_0}{\partial x}\right) + B_{16}\frac{\partial \phi_x}{\partial x} + B_{26}\frac{\partial \phi_y}{\partial y}\right. \\ &\quad \left. + B_{66}\left(\frac{\partial \phi_x}{\partial y} + \frac{\partial \phi_y}{\partial x}\right)\right] + \frac{\partial}{\partial y}\left[A_{12}\frac{\partial u_0}{\partial x} + A_{22}\frac{\partial v_0}{\partial y}\right. \\ &\quad \left. + A_{26}\left(\frac{\partial u_0}{\partial y} + \frac{\partial v_0}{\partial x}\right) + B_{12}\frac{\partial \phi_x}{\partial x} + B_{22}\frac{\partial \phi_y}{\partial y} + B_{26}\left(\frac{\partial \phi_x}{\partial y} + \frac{\partial \phi_y}{\partial x}\right)\right] \\ &\quad - \left(\frac{\partial N_{xy}^T}{\partial x} + \frac{\partial N_{yy}^T}{\partial y}\right) = I_0\frac{\partial^2 v_0}{\partial t^2} + I_1\frac{\partial^2 \phi_y}{\partial t^2} \end{aligned} \qquad (4.3-15)$$

$$\begin{aligned}&\frac{\partial}{\partial x}\left[KA_{45}\left(\frac{\partial w_0}{\partial y} + \phi_y\right) + KA_{55}\left(\frac{\partial w_0}{\partial x} + \phi_x\right)\right] \\ &\quad + \frac{\partial}{\partial y}\left[KA_{44}\left(\frac{\partial w_0}{\partial y} + \phi_y\right) + KA_{45}\left(\frac{\partial w_0}{\partial x} + \phi_x\right)\right] \\ &\quad + \hat{N}_{xx}\frac{\partial^2 w_0}{\partial x^2} + 2\hat{N}_{xy}\frac{\partial^2 w_0}{\partial x \partial y} + \hat{N}_{yy}\frac{\partial^2 w_0}{\partial y^2} + q = I_0\frac{\partial^2 w_0}{\partial t^2} \end{aligned} \qquad (4.3-16)$$

$$\begin{aligned}
&\frac{\partial}{\partial x}\left[B_{11}\frac{\partial u_0}{\partial x} + B_{12}\frac{\partial v_0}{\partial y} + B_{16}\left(\frac{\partial u_0}{\partial y} + \frac{\partial v_0}{\partial x}\right) + D_{11}\frac{\partial \phi_x}{\partial x} + D_{12}\frac{\partial \phi_y}{\partial y}\right.\\
&\quad\left. + D_{16}\left(\frac{\partial \phi_x}{\partial y} + \frac{\partial \phi_y}{\partial x}\right)\right] + \frac{\partial}{\partial y}\left[B_{16}\frac{\partial u_0}{\partial x} + B_{26}\frac{\partial v_0}{\partial y}\right.\\
&\quad\left. + B_{66}\left(\frac{\partial u_0}{\partial y} + \frac{\partial v_0}{\partial x}\right) + D_{16}\frac{\partial \phi_x}{\partial x} + D_{26}\frac{\partial \phi_y}{\partial y} + D_{66}\left(\frac{\partial \phi_x}{\partial y} + \frac{\partial \phi_y}{\partial x}\right)\right]\\
&\quad - \left[KA_{45}\left(\frac{\partial w_0}{\partial y} + \phi_y\right) + KA_{55}\left(\frac{\partial w_0}{\partial x} + \phi_x\right)\right]\\
&\quad - \left(\frac{\partial M_{xx}^T}{\partial x} + \frac{\partial M_{xy}^T}{\partial y}\right) = I_0\frac{\partial^2 \phi_x}{\partial t^2} + I_1\frac{\partial^2 u_0}{\partial t^2} \qquad (4.3-17)
\end{aligned}$$

$$\begin{aligned}
&\frac{\partial}{\partial x}\left[B_{16}\frac{\partial u_0}{\partial x} + B_{26}\frac{\partial v_0}{\partial y} + B_{66}\left(\frac{\partial u_0}{\partial y} + \frac{\partial v_0}{\partial x}\right) + D_{16}\frac{\partial \phi_x}{\partial x}\right.\\
&\quad\left. + D_{26}\frac{\partial \phi_y}{\partial y} + D_{66}\left(\frac{\partial \phi_x}{\partial y} + \frac{\partial \phi_y}{\partial x}\right)\right] + \frac{\partial}{\partial y}\left[B_{12}\frac{\partial u_0}{\partial x} + B_{22}\frac{\partial v_0}{\partial y}\right.\\
&\quad\left. + B_{26}\left(\frac{\partial u_0}{\partial y} + \frac{\partial v_0}{\partial x}\right) + D_{12}\frac{\partial \phi_x}{\partial y} + D_{22}\frac{\partial \phi_y}{\partial y} + D_{26}\left(\frac{\partial \phi_x}{\partial y} + \frac{\partial \phi_y}{\partial x}\right)\right]\\
&\quad - \left[KA_{44}\left(\frac{\partial w_0}{\partial y} + \phi_y\right) + KA_{45}\left(\frac{\partial w_0}{\partial x} + \phi_x\right)\right]\\
&\quad - \left(\frac{\partial M_{xy}^T}{\partial x} + \frac{\partial M_{yy}^T}{\partial y}\right) = I_0\frac{\partial^2 \phi_y}{\partial t^2} + I_1\frac{\partial^2 v_0}{\partial t^2} \qquad (4.3-18)
\end{aligned}$$

This completes the development of the governing equations of the first–order shear deformation laminate theory. We will return to these equations in the next chapter to develop their solutions for specific plate problems.

4.4 STIFFNESS CHARACTERISTICS OF SELECTED LAMINATION SCHEMES

4.4–1 Terminology and Notation

It is useful to review the terminology and notation used in connection with lamination schemes or stacking sequences (see Reddy [15] for additional details). The lamination scheme of a laminate will be denoted by $[\alpha/\beta/90/-\alpha/0/\cdots]$, where α is the orientation of the first ply, β is the orientation of the second ply, and so on. The plies are counted in the positive z direction (see Fig. 4.2–1). Unless stated otherwise, this notation also implies that all layers are of the same thickness and made of the same material.

A general laminate has layers of different orientations θ where $-90^\circ \leq \theta \leq 90^\circ$. For example, [0/15/-35/45/90/-45] is a six–ply laminate. *Angle–ply* laminates have ply orientations of θ or $-\theta$ where $0^\circ \leq \theta \leq 90^\circ$, with at least one layer having an orientation other than 0° 90°. An example of angle–ply laminates is provided by [15/ – 30/0/90/45/ – 45]. *Cross–ply* laminates are those which have ply orientations of 0° or ±90°. An example of a cross–ply laminate is [0/90/ – 90/0/0/90].

When ply stacking sequence, material, and geometry (*i.e.*, ply thicknesses) are symmetric about the midplane of the laminate, the laminate is called a *symmetric laminate.* For a symmetric laminate, the upper half through the laminate thickness is a mirror image of the lower half. The laminates [–45/45/45/ – 45] and [45/ – 45/ – 45/45], with all layers having the same thickness, are examples of a symmetric angle–ply laminate, [0/90/90/0] is a symmetric cross–ply laminate, and [30/ – 45/0/90/90/0/ – 45/30] is a general symmetric laminate. Symmetric laminates are also denoted $[\ ]_s$ by displaying only the lamination scheme of the upper half. The symmetric laminate [30/ – 45/0/90/90/0/ – 45/30] can be denoted as $[30/-45/0/90]_s$. An unsymmetric or *asymmetric laminate* is a laminate that is not symmetric. An *antisymmetric* laminate is one whose lamination scheme, material, and geometry are antisymmetric about the midplane. Examples of antisymmetric angle–ply and cross–ply laminates are provided, respectively, by [–30/30/ – 30/30/ – 30/30] and [0/90/0/90/0/90].

4.4–2 Single–Layer, Generally Orthotropic, Plates

For a single generally orthotropic layer (*i.e.*, the principal material coordinates do not coincide with those of the plate), we have $B_{ij} = 0$, and

$$A_{ij} = \bar{Q}_{ij}h \ , \quad D_{ij} = \frac{\bar{Q}_{ij}h^3}{12} \ , \quad A_{44} = KhQ_{44}, \ A_{55} = KhQ_{55} \tag{4.4 – 1}$$

The single–layer plate constitutive equations are

$$\begin{Bmatrix} N_{xx} \\ N_{yy} \\ N_{xy} \end{Bmatrix} = \begin{bmatrix} A_{11} & A_{12} & A_{16} \\ A_{12} & A_{22} & A_{26} \\ A_{16} & A_{26} & A_{66} \end{bmatrix} \begin{Bmatrix} \varepsilon_{xx}^{(0)} \\ \varepsilon_{yy}^{(0)} \\ \gamma_{xy}^{(0)} \end{Bmatrix} \qquad (4.4-2)$$

$$\begin{Bmatrix} M_{xx} \\ M_{yy} \\ M_{xy} \end{Bmatrix} = \begin{bmatrix} D_{11} & D_{12} & D_{16} \\ D_{12} & D_{22} & D_{26} \\ D_{16} & D_{26} & D_{66} \end{bmatrix} \begin{Bmatrix} \varepsilon_{xx}^{(1)} \\ \varepsilon_{yy}^{(1)} \\ \gamma_{xy}^{(1)} \end{Bmatrix} \qquad (4.4-3)$$

$$\begin{Bmatrix} Q_{y} \\ Q_{x} \end{Bmatrix} = K \begin{bmatrix} A_{44} & A_{45} \\ A_{45} & A_{55} \end{bmatrix} \begin{Bmatrix} \gamma_{yz}^{(0)} \\ \gamma_{xz}^{(0)} \end{Bmatrix} \qquad (4.4-4)$$

If the temperature increment is linear through the layer thickness, $\Delta T = T_0 + zT_1$, the thermal stress resultants have the form

$$\begin{Bmatrix} N_{xx}^T \\ N_{yy}^T \\ N_{xy}^T \end{Bmatrix} = \begin{bmatrix} \bar{Q}_{11} & \bar{Q}_{12} & \bar{Q}_{16} \\ \bar{Q}_{12} & \bar{Q}_{22} & \bar{Q}_{26} \\ \bar{Q}_{16} & \bar{Q}_{26} & \bar{Q}_{66} \end{bmatrix} \begin{Bmatrix} \alpha_{xx} \\ \alpha_{yy} \\ \alpha_{xy} \end{Bmatrix} T_0 h \qquad (4.4-5a)$$

$$\begin{Bmatrix} M_{xx}^T \\ M_{yy}^T \\ M_{xy}^T \end{Bmatrix} = \begin{bmatrix} \bar{Q}_{11} & \bar{Q}_{12} & \bar{Q}_{16} \\ \bar{Q}_{12} & \bar{Q}_{22} & \bar{Q}_{26} \\ \bar{Q}_{16} & \bar{Q}_{26} & \bar{Q}_{66} \end{bmatrix} \begin{Bmatrix} \alpha_{xx} \\ \alpha_{yy} \\ \alpha_{xy} \end{Bmatrix} \frac{T_1 h^3}{12} \qquad (4.4-5b)$$

4.4–3 Symmetric Laminates

For symmetric laminates (*i.e.*, when the material properties, locations, and lamination scheme are symmetric about the midplane of the laminate), the coupling stiffnesses B_{ij} are zero. The elimination of the coupling between bending and extension simplifies the governing equations. For example, the governing equations (4.2–21) and (4.2–22) of symmetric laminates according to the classical laminate theory become uncoupled from Eq. (4.2–23) when the extensional–bending coupling stiffnesses B_{ij} are zero. In that case Eqs. (4.2–21) and (4.2–22) can be solved independently for the inplane displacements (u_0, v_0), and Eq. (4.2–23) can be solved for the transverse deflection w_0. For the isothermal case, the transverse deflection w_0 under applied transverse force q is given by

$$
\begin{aligned}
&-\left[D_{11}\frac{\partial^4 w_0}{\partial x^4} + 2(D_{12}+2D_{66})\frac{\partial^4 w_0}{\partial x^2 \partial y^2} + D_{22}\frac{\partial^4 w_0}{\partial y^4}\right] \\
&-4\left(D_{16}\frac{\partial^4 w_0}{\partial x^3 \partial y} + D_{26}\frac{\partial^4 w_0}{\partial x \partial y^3}\right) + q \\
&+\hat{N}_{xx}\frac{\partial^2 w_0}{\partial x^2} + 2\hat{N}_{xy}\frac{\partial^2 w_0}{\partial x \partial y} + \hat{N}_{yy}\frac{\partial^2 w_0}{\partial y^2} = I_0\frac{\partial^2 w_0}{\partial t^2} \qquad (4.4-6)
\end{aligned}
$$

In the case of the first–order theory, Eqs. (4.3–16)–(4.3–18) involving (w_0, ϕ_x, ϕ_y) can be solved independent of Eqs. (4.3–14) and (4.3–15). Equations (4.3–17) and (4.3–18) become

$$
\begin{aligned}
&D_{11}\frac{\partial^2 \phi_x}{\partial x^2} + D_{16}\frac{\partial^2 \phi_y}{\partial x^2} + (D_{12}+D_{66})\frac{\partial^2 \phi_y}{\partial x \partial y} + 2D_{16}\frac{\partial^2 \phi_x}{\partial x \partial y} + D_{66}\frac{\partial^2 \phi_x}{\partial y^2} \\
&+ D_{26}\frac{\partial^2 \phi_y}{\partial y^2} - A_{45}\left(\phi_y + \frac{\partial w_0}{\partial y}\right) - A_{55}\left(\phi_x + \frac{\partial w_0}{\partial x}\right) \\
&-\left(\frac{\partial M_{xx}^T}{\partial x} + \frac{\partial M_{xy}^T}{\partial y}\right) = I_0\frac{\partial^2 \phi_x}{\partial t^2} \qquad (4.4-7)
\end{aligned}
$$

$$
\begin{aligned}
&D_{16}\frac{\partial^2 \phi_x}{\partial x^2} + D_{66}\frac{\partial^2 \phi_y}{\partial x^2} + (D_{12}+D_{66})\frac{\partial^2 \phi_x}{\partial x \partial y} + 2D_{26}\frac{\partial^2 \phi_y}{\partial x \partial y} + D_{26}\frac{\partial^2 \phi_x}{\partial y^2} \\
&+ D_{22}\frac{\partial^2 \phi_y}{\partial y^2} - A_{44}\left(\phi_y + \frac{\partial w_0}{\partial y}\right) - A_{45}\left(\phi_x + \frac{\partial w_0}{\partial x}\right) \\
&-\left(\frac{\partial M_{xy}^T}{\partial x} + \frac{\partial M_{yy}^T}{\partial y}\right) = I_0\frac{\partial^2 \phi_y}{\partial t^2} \qquad (4.4-8)
\end{aligned}
$$

4.4–4 Antisymmetric Laminates

Although symmetric laminates are more desirable from an analysis standpoint, they may not meet the design requirements in some applications. For example, a heat shield receives heat from one side, and thus requires nonsymmetric laminates to shield the heat effectively. Another example that requires coupling is provided by turbine blades with pretwist. Moreover, the shear stiffness of laminates can be increased by orienting the layers at an angle to the laminate coordinates.

The general class of antisymmetric laminates must have an *even* number of orthotropic laminae if adjacent laminae have equal thicknesses and alternating orientations: $\theta/-\theta, 0^\circ \leq \theta \leq 90^\circ$. Due to the antisymmetry of the lamination scheme but symmetry of the thicknesses of each pair of layers, this class of antisymmetric laminates has the feature that $A_{16} = A_{26} = D_{16} = D_{26} = 0$. The coupling stiffnesses B_{ij} are not zero. The relations between the stress resultants and the strains are

$$\left\{ \begin{matrix} N_{xx} \\ N_{yy} \\ N_{xy} \end{matrix} \right\} = \begin{bmatrix} A_{11} & A_{12} & 0 \\ A_{12} & A_{22} & 0 \\ 0 & 0 & A_{66} \end{bmatrix} \left\{ \begin{matrix} \varepsilon_{xx}^{(0)} \\ \varepsilon_{yy}^{(0)} \\ \gamma_{xy}^{(0)} \end{matrix} \right\} + \begin{bmatrix} B_{11} & B_{12} & B_{16} \\ B_{12} & B_{22} & B_{26} \\ B_{16} & B_{26} & B_{66} \end{bmatrix} \left\{ \begin{matrix} \varepsilon_{xx}^{(1)} \\ \varepsilon_{yy}^{(1)} \\ \gamma_{xy}^{(1)} \end{matrix} \right\} \tag{4.4-10}$$

$$\left\{ \begin{matrix} M_{xx} \\ M_{yy} \\ M_{xy} \end{matrix} \right\} = \begin{bmatrix} B_{11} & B_{12} & B_{16} \\ B_{12} & B_{22} & B_{26} \\ B_{16} & B_{26} & B_{66} \end{bmatrix} \left\{ \begin{matrix} \varepsilon_{xx}^{(0)} \\ \varepsilon_{yy}^{(0)} \\ \gamma_{xy}^{(0)} \end{matrix} \right\} + \begin{bmatrix} D_{11} & D_{12} & 0 \\ D_{12} & D_{22} & 0 \\ 0 & 0 & D_{66} \end{bmatrix} \left\{ \begin{matrix} \varepsilon_{xx}^{(1)} \\ \varepsilon_{yy}^{(1)} \\ \gamma_{xy}^{(1)} \end{matrix} \right\} \tag{4.4-11}$$

$$\left\{ \begin{matrix} Q_y \\ Q_x \end{matrix} \right\} = \begin{bmatrix} A_{44} & 0 \\ 0 & A_{55} \end{bmatrix} \left\{ \begin{matrix} \gamma_{yz}^{(0)} \\ \gamma_{xz}^{(0)} \end{matrix} \right\} \tag{4.4-12}$$

The thermal force resultants are given by

$$\begin{Bmatrix} N_{xx}^T \\ N_{yy}^T \\ N_{xy}^T \end{Bmatrix} = \sum_{k=1}^{L} \int_{z_k}^{z_{k+1}} \begin{bmatrix} \bar{Q}_{11} & \bar{Q}_{12} & \bar{Q}_{16} \\ \bar{Q}_{12} & \bar{Q}_{22} & \bar{Q}_{26} \\ \bar{Q}_{16} & \bar{Q}_{26} & \bar{Q}_{66} \end{bmatrix}^{(k)} \begin{Bmatrix} \alpha_{xx} \\ \alpha_{yy} \\ \alpha_{xy} \end{Bmatrix}^{(k)} \Delta T dz \tag{4.4-13}$$

A similar expression holds for $\{M^T\}$.

Antisymmetric Cross–Ply Laminates

A special case of antisymmetric laminates are those that have an even number of orthotropic layers with principal material directions alternating at 0° to 90° to the laminate axes. Such laminates are called *antisymmetric cross–ply laminates.* Examples of antisymmetric cross–ply laminates are: [0/90/0/90/ $\cdots$] with all layers of the same thickness, and [0/90/90/0/0/90] with layers of the following thicknesses $[h_1/h_2/h_3/h_3/h_2/h_1]$. Note that for every 0° layer of a given thickness and location, there is a 90° layer of the same thickness and location on the other side of the midplane. For these laminates, the coupling stiffnesses B_{ij} have the properties

$$B_{22} = -B_{11}, \quad \text{and all other} \quad B_{ij} = 0 \tag{4.4-14}$$

The relations between the stress resultants and the strains are

$$\begin{Bmatrix} N_{xx} \\ N_{yy} \\ N_{xy} \end{Bmatrix} = \begin{bmatrix} A_{11} & A_{12} & 0 \\ A_{12} & A_{22} & 0 \\ 0 & 0 & A_{66} \end{bmatrix} \begin{Bmatrix} \varepsilon_{xx}^{(0)} \\ \varepsilon_{yy}^{(0)} \\ \gamma_{xy}^{(0)} \end{Bmatrix} + \begin{bmatrix} B_{11} & 0 & 0 \\ 0 & -B_{11} & 0 \\ 0 & 0 & 0 \end{bmatrix} \begin{Bmatrix} \varepsilon_{xx}^{(1)} \\ \varepsilon_{yy}^{(1)} \\ \gamma_{xy}^{(1)} \end{Bmatrix} \tag{4.4-15}$$

$$\begin{Bmatrix} M_{xx} \\ M_{yy} \\ M_{xy} \end{Bmatrix} = \begin{bmatrix} B_{11} & 0 & 0 \\ 0 & -B_{11} & 0 \\ 0 & 0 & 0 \end{bmatrix} \begin{Bmatrix} \varepsilon_{xx}^{(0)} \\ \varepsilon_{yy}^{(0)} \\ \gamma_{xy}^{(0)} \end{Bmatrix} + \begin{bmatrix} D_{11} & D_{12} & 0 \\ D_{12} & D_{22} & 0 \\ 0 & 0 & D_{66} \end{bmatrix} \begin{Bmatrix} \varepsilon_{xx}^{(1)} \\ \varepsilon_{yy}^{(1)} \\ \gamma_{xy}^{(1)} \end{Bmatrix} \tag{4.4 – 16}$$

$$\begin{Bmatrix} Q_y \\ Q_x \end{Bmatrix} = \begin{bmatrix} A_{44} & 0 \\ 0 & A_{55} \end{bmatrix} \begin{Bmatrix} \gamma_{yz}^{(0)} \\ \gamma_{xz}^{(0)} \end{Bmatrix} \tag{4.4 – 17}$$

A *regular antisymmetric cross–ply laminate* is one that has an even number of layers of equal thickness and material properties and have alternating 0° and 90° orientations. For these laminates, the coupling coefficient B_{11} can be shown to approach zero as the number of layers increases.

Antisymmetric Angle–Ply Laminates

An antisymmetric angle–ply laminate has an even number of orthotropic layers with principal material directions alternating at θ degrees to the laminate axes on one side of the midplane and corresponding equal thickness laminae oriented at $-\theta$ degrees on the other side. A *regular antisymmetric angle–ply laminate* is one that has an even number of layers of equal thickness and material properties. An example is given by $[-45/40/-15/15/-40/45]$.

For antisymmetric angle–ply laminates the stiffnesses can be simplified as

$$A_{16} = A_{26} = D_{16} = D_{26} = 0; \quad B_{11} = B_{22} = B_{12} = B_{66} = 0 \tag{4.4 – 18}$$

The relations between the stress resultants and the strains are

$$\begin{Bmatrix} N_{xx} \\ N_{yy} \\ N_{xy} \end{Bmatrix} = \begin{bmatrix} A_{11} & A_{12} & 0 \\ A_{12} & A_{22} & 0 \\ 0 & 0 & A_{66} \end{bmatrix} \begin{Bmatrix} \varepsilon_{xx}^{(0)} \\ \varepsilon_{yy}^{(0)} \\ \gamma_{xy}^{(0)} \end{Bmatrix} + \begin{bmatrix} 0 & 0 & B_{16} \\ 0 & 0 & B_{26} \\ B_{16} & B_{26} & 0 \end{bmatrix} \begin{Bmatrix} \varepsilon_{xx}^{(1)} \\ \varepsilon_{yy}^{(1)} \\ \gamma_{xy}^{(1)} \end{Bmatrix} \tag{4.4-19}$$

$$\begin{Bmatrix} M_{xx} \\ M_{yy} \\ M_{xy} \end{Bmatrix} = \begin{bmatrix} 0 & 0 & B_{16} \\ 0 & 0 & B_{26} \\ B_{16} & B_{26} & 0 \end{bmatrix} \begin{Bmatrix} \varepsilon_{xx}^{(0)} \\ \varepsilon_{yy}^{(0)} \\ \gamma_{xy}^{(0)} \end{Bmatrix} + \begin{bmatrix} D_{11} & D_{12} & 0 \\ D_{12} & D_{22} & 0 \\ 0 & 0 & D_{66} \end{bmatrix} \begin{Bmatrix} \varepsilon_{xx}^{(1)} \\ \varepsilon_{yy}^{(1)} \\ \gamma_{xy}^{(1)} \end{Bmatrix} \tag{4.4-20}$$

$$\begin{Bmatrix} Q_y \\ Q_x \end{Bmatrix} = \begin{bmatrix} A_{44} & 0 \\ 0 & A_{55} \end{bmatrix} \begin{Bmatrix} \gamma_{yz}^{(0)} \\ \gamma_{xz}^{(0)} \end{Bmatrix} \tag{4.4-21}$$

For a fixed laminate thickness, the stiffnesses B_{16} and B_{26} go to zero as the number of layers in the laminate increases.

EXERCISE PROBLEMS

4.1 *The Euler–Bernoulli Beam Theory.* Let the displacement field of a beam in bending be

$$\begin{aligned} u_1(x,z,t) &= u_0(x,t) - z\frac{\partial w_0}{\partial x}(x,t) \\ u_2(x,z,t) &= 0 \\ u_3(x,z,t) &= w_0(x,t) \end{aligned}$$

Compute the linear strains associated with the displacement field.

4.2 *The Timoshenko Beam Theory.* Suppose that the displacement field of the Timoshenko beam theory is

$$
\begin{aligned}
u_1(x,z,t) &= u_0(x,t) + z\phi(x,t) \\
u_2(x,z,t) &= 0 \\
u_3(x,z,t) &= w_0(x,t)
\end{aligned}
$$

Compute the linear strains associated with the displacement field.

4.3 Using the principle of virtual displacements

$$
\begin{aligned}
\delta W &\equiv \delta W_I + \delta W_E \\
&= \int_0^L \int_A \sigma_{xx} \delta\varepsilon_{xx} dA \; dx - \int_0^L q\delta w_0 dx
\end{aligned}
$$

derive the equations of equilibrium of the classical laminated beam theory.

4.4 Repeat Problem 3 for the case of the Timoshenko beam theory. Note that the virtual work expression for this case is given by

$$
\begin{aligned}
\delta W &\equiv \delta W_I + \delta W_E \\
&= \int_0^L \int_A (\sigma_{xx}\delta\varepsilon_{xx} + 2\sigma_{xz}\delta\varepsilon_{xz}) \, dA \; dx - \int_0^L q\delta w_0 dx
\end{aligned}
$$

4.5 Using the principle of virtual displacements

$$
\begin{aligned}
\delta W &\equiv \delta W_I + \delta W_E \\
&= \int_\omega \int_{\frac{h}{2}}^{\frac{h}{2}} \sigma_{ij}\delta\varepsilon_{ij} dz \; dxdy - \int_\omega q\delta w_0 dxdy
\end{aligned}
$$

derive the equations of equilibrium of the classical laminated plate theory. Note that the virtual strains are given by [see Eqs. (4.2–4)]

$$\begin{aligned}
\delta\varepsilon_{xx} &= \frac{\partial \delta u_0}{\partial x} - z\frac{\partial^2 \delta w_0}{\partial x^2} \\
\delta\varepsilon_{yy} &= \frac{\partial \delta v_0}{\partial y} - z\frac{\partial^2 \delta w_0}{\partial y^2} \\
2\delta\varepsilon_{xy} &= \frac{\partial \delta u_0}{\partial y} + \frac{\partial \delta v_0}{\partial x} - 2z\frac{\partial^2 \delta w_0}{\partial x \partial y}
\end{aligned}$$

4.6 Repeat Problem 5 for the first–order shear deformation theory.

4.7 Show that the membrane strains $\{\varepsilon^0\}$ and the moment resultants $\{M\}$ in the classical or first–order laminated plate theory can be expressed in terms of force resultants $\{N\}$ and bending strains $\{\varepsilon^1\}$ as

$$\begin{aligned}
\{\varepsilon^0\} &= [A]^{-1}\left(\{N\} - [B]\{\varepsilon^1\}\right) \\
\{M\} &= \left([B][A]^{-1}\right)\{N\} - \left([B][A]^{-1}[B] - [D]\right)\{\varepsilon^1\}
\end{aligned}$$

These equations bring out the bending–extensional coupling for laminates with nonzero $[B]$. For example, when the bending strains are zero, the applied inplane forces induce bending moments for laminates with nonzero coupling coefficients $[B]$.

4.8 Show that for a general laminate composed of multiple isotropic layers, the laminate stiffness $A_{16}, A_{26}, B_{16}, B_{26}, D_{16}$ and D_{26} are zero, and that $A_{22} = A_{11}, B_{22} = B_{11}$ and $D_{22} = D_{11}$.

4.9 Show that for a general laminate composed of multiple specially orthotropic layers, the laminate stiffness $A_{16}, A_{26}, B_{16}, B_{26}, D_{16}$ and D_{26} are zero.

4.10 **(Laminated Beam)** Consider a symmetrically laminated beam. Assuming that the thermal forces and moments are zero and

$$M_{yy} = M_{xy} = 0 \tag{a}$$

everywhere in the beam, derive the equation governing the transverse deflection of the beam according to the classical plate theory; *i.e.*, use Eq. (4.2–18) to express the curvatures in terms of the bending moments

$$\frac{\partial^2 w_0}{\partial x^2} = -D_{11}^* M_{xx} \tag{b}$$

where

$$\begin{aligned} D_{11}^* &= (D_{22}D_{66} - D_{26}D_{26})/D^* \\ D^* &= (D_{11}D_{22} - D_{12}D_{12})D_{66} + (D_{12}D_{16} - D_{11}D_{26})D_{26} \\ &\quad + (D_{12}D_{26} - D_{16}D_{22})D_{16} \end{aligned} \tag{c}$$

Recast Eq. (b) in the familiar form of the classical Euler–Bernoulli beam theory by introducing the quantities

$$M = bM_{xx} \ , \quad E_{xx}^b = \frac{12}{h^3 D_{11}^*} = \frac{b}{I_{yy} D_{11}^*} \ , \quad I_{yy} = \frac{bh^3}{12} \tag{d}$$

Show that

$$\frac{\partial^2 w_0}{\partial x^2} = -\frac{M}{E_{xx}^b I_{yy}} \tag{e}$$

where b is the width, and h is the total thickness of the laminated beam.

4.11 Assume that $M_{yy} = M_{xy} = Q_y = \phi_y = 0$ and both w_0 and ϕ_x are functions of only x and t in the first–order shear deformation plate theory. Show that

$$\frac{\partial \phi_x}{\partial x} = D_{11}^* M_{xx} \ , \quad \frac{\partial w_0}{\partial x} + \phi_x = \frac{A_{55}^*}{K} Q_x \tag{a}$$

where D_{11}^* is as defined in Problem 4.10, and

$$A_{55}^* = \frac{A_{44}}{A} \ , \quad A = A_{44}A_{55} - A_{45}A_{45} \tag{b}$$

Introduce the variables

$$M(x) = bM_{xx} \ , \quad Q(x) = bQ_x \ , \quad E_{xx}^b = \frac{12}{D_{11}^* h^3} \ , \quad G_{xz}^b = \frac{1}{A_{55}^* h} \tag{c}$$

where K denotes the the shear correction coefficient, and obtain the following governing equations of the laminated beam according to the Timoshenko beam theory

$$E_{xx}^b I_{yy} \frac{\partial \phi_x}{\partial x} = M(x) \tag{d}$$

$$KG_{xz}^b bh \left(\frac{\partial w_0}{\partial x} + \phi_x \right) = Q(x) \tag{e}$$

4.12 **(Cylindrical Bending)** Consider a plate strip that has a finite dimension along the $x-$axis and subjected to a transverse load $q(x)$ that is uniform at any section parallel to the $x-$axis. In such a case, the deflection w_0 and displacements (u_0, v_0) of the plate are functions of only x. Therefore, all derivatives with respect to y are zero. In such cases, the deflected surface of the plate strip is cylindrical, and it is referred to as the *cylindrical bending*. For this case, show that the governing equations according to the classical laminate theory are

$$A_{11} \left(\frac{\partial^2 u_0}{\partial x^2} + \frac{\partial w_0}{\partial x} \frac{\partial^2 w_0}{\partial x^2} \right) + A_{16} \frac{\partial^2 v_0}{\partial x^2} - B_{11} \frac{\partial^3 w_0}{\partial x^3} - \frac{\partial N_{xx}^T}{\partial x}$$
$$= I_0 \frac{\partial^2 u_0}{\partial t^2} - I_1 \frac{\partial^3 w_0}{\partial x \partial t^2} \tag{a}$$

$$A_{16} \left(\frac{\partial^2 u_0}{\partial x^2} + \frac{\partial w_0}{\partial x} \frac{\partial^2 w_0}{\partial x^2} \right) + A_{66} \frac{\partial^2 v_0}{\partial x^2} - B_{16} \frac{\partial^3 w_0}{\partial x^3} - \frac{\partial N_{xy}^T}{\partial x}$$
$$= I_0 \frac{\partial^2 v_0}{\partial t^2} \tag{b}$$

$$B_{11}\left(\frac{\partial^3 u_0}{\partial x^3}+\frac{\partial^2 w_0}{\partial x^2}\frac{\partial^2 w_0}{\partial x^2}+\frac{\partial w_0}{\partial x}\frac{\partial^3 w_0}{\partial x^3}\right)+B_{16}\frac{\partial^3 v_0}{\partial x^3}-D_{11}\frac{\partial^4 w_0}{\partial x^4}$$
$$+q-\frac{\partial^2 M_{xx}^T}{\partial x^2}=I_0\frac{\partial^2 w_0}{\partial t^2}-I_2\frac{\partial^4 w_0}{\partial x^2 \partial t^2}+I_1\frac{\partial^3 u_0}{\partial x \partial t^2} \qquad (c)$$

4.13 Show that the equations of motion for cylindrical bending according to the first–order shear deformation theory are given (see Problem 4.12) by

$$A_{11}\frac{\partial^2 u_0}{\partial x^2}+A_{16}\frac{\partial^2 v_0}{\partial x^2}+B_{11}\frac{\partial^2 \phi_x}{\partial x^2}+B_{16}\frac{\partial^2 \phi_y}{\partial x^2}-\frac{\partial N_{xx}^T}{\partial x}$$
$$=I_0\frac{\partial^2 u_0}{\partial t^2}+I_1\frac{\partial^2 \phi_x}{\partial t^2} \qquad (a)$$

$$A_{16}\frac{\partial^2 u_0}{\partial x^2}+A_{66}\frac{\partial^2 v_0}{\partial x^2}+B_{16}\frac{\partial^2 \phi_x}{\partial x^2}+B_{66}\frac{\partial^2 \phi_y}{\partial x^2}-\frac{\partial N_{xy}^T}{\partial x}$$
$$=I_0\frac{\partial^2 v_0}{\partial t^2}+I_1\frac{\partial^2 \phi_y}{\partial t^2} \qquad (b)$$

$$B_{11}\frac{\partial^2 u_0}{\partial x^2}+B_{16}\frac{\partial^2 v_0}{\partial x^2}+D_{11}\frac{\partial^2 \phi_x}{\partial x^2}+D_{16}\frac{\partial^2 \phi_y}{\partial x^2}-KA_{55}\left(\frac{\partial w_0}{\partial x}+\phi_x\right)$$
$$-KA_{45}\phi_y-\frac{\partial M_{xx}^T}{\partial x}=I_2\frac{\partial^2 \phi_x}{\partial t^2}+I_1\frac{\partial^2 u_0}{\partial t^2} \qquad (c)$$

$$B_{16}\frac{\partial^2 u_0}{\partial x^2}+B_{66}\frac{\partial^2 v_0}{\partial x^2}+D_{16}\frac{\partial^2 \phi_x}{\partial x^2}+D_{66}\frac{\partial^2 \phi_y}{\partial x^2}-KA_{44}\phi_y$$
$$-KA_{45}\left(\frac{\partial w_0}{\partial x}+\phi_x\right)-\frac{\partial M_{xy}^T}{\partial x}=I_2\frac{\partial^2 \phi_y}{\partial t^2}+I_1\frac{\partial^2 v_0}{\partial t^2} \qquad (d)$$

$$KA_{55}\left(\frac{\partial^2 w_0}{\partial x^2}+\frac{\partial \phi_x}{\partial x}\right)+KA_{45}\frac{\partial \phi_y}{\partial x}+q=I_0\frac{\partial^2 w_0}{\partial t^2} \qquad (e)$$

where K is the shear correction coefficient, and I_0 and I_1 are the mass inertias.

REFERENCES FOR ADDITIONAL READING

1. Reddy, J. N., *Energy and Variational Methods in Applied Mechanics*, John Wiley & Sons, New York, 1984.

2. Reddy, J. N., *Mechanics of Laminated Composite Structures: Theory and Analysis,* Lecture Notes, Virginia Polytechnic Institute and State University, Blacksburg, VA, October 1988; also see Chapters 14 and 15 in: *Finite Element Analysis for Engineering Design*, J. N. Reddy, C. S. Krishna Moorthy, and K. N. Seetharamu (eds.), Vol. 37, Springer–Verlag, Berlin, 1988.

3. Reddy, J. N., "A Review of Refined Theories of Laminated Plates," *Shock and Vibration Digest*, **22** (7), pp. 3–17, 1990.

4. Reddy, J. N., "A General Non–Linear Third–Order Theory of Plates with Transverse Shear Deformation," *Journal of Non–Linear Mechanics*, **25** (6), pp. 677–686, 1990.

5. Reddy, J. N., "Generalization of Two–Dimensional Theories of Laminated Composite Plates," *Communications in Applied Numerical Methods,* **3**, pp. 173–180, 1987.

6. Robbins, D. H. and Reddy, J. N., "Modeling of Thick Composites Using a Layerwise Laminate Theory," *International Journal for Numerical Methods in Engineering,* **36**, 655–677, 1993.

7. Reddy, J. N. and D. H. Robbins, Jr., "On Structural Theories and Computational Models for Composite Laminates," *Applied Mechanics Reviews,* **47** (6), Part 1, pp. 147–170, 1994.

8. Mindlin, R. D., "Influence of Rotatory Inertia and Shear on Flexural Motions of Isotropic, Elastic Plates," *Journal of Applied Mechanics,* **18**, pp. 31–38, 1951.

9. Basset, A. B., "On the Extension and Flexure of Cylindrical and Spherical Thin Elastic Shells," *Philosophical Transactions of the Royal Society of London,* Ser. A, **181** (6), pp. 433–480, 1890.

10. Hildebrand, F. B., Reissner, E., and Thomas, G. B., "Notes on the Foundations of the Theory of Small Displacements of Orthotropic Shells," NASA TN–1833, Washington, D. C., 1949.

11. Hencky, H., "Uber die Berucksichtigung der Schubverzerrung in ebenen Platten," *Inginier Archive,* **16**, pp. 72–76, 1947.

12. Chatterjee, S. N. and Kulkarni, S. V., "Shear Correction Factors for Laminated Plates," *AIAA Journal,* **17** (5), pp. 498–499, 1979.

13. Whitney, J. M., "Shear Correction Factors for Orthotropic Laminates Under Static Load," *Journal of Applied Mechanics,* **40** (1), pp. 302–304, 1973.

14. Bert, C. W., "Simplified Analysis of Static Shear Factors for Beams of Nonhomogeneous Cross Section," *Journal of Composite Materials*, **7**, pp. 525–529, 1973.

15. Reddy, J. N., *Mechanics of Laminated Composite Structures: Theory and Analysis,* John Wiley & Sons, New York, 1996, in press.

Chapter Five

ANALYSIS OF RECTANGULAR PLATES

5.1 INTRODUCTION

The partial differential equations governing composite laminates (see Section 4.3) of arbitrary geometries and boundary conditions cannot be solved in closed form. Analytical solutions of plate theories are available (see Reddy [4–8]) mostly for rectangular plates with all edges simply supported (*i.e.*, the Navier solutions) or with two opposite edges simply supported and the remaining edges having arbitrary boundary conditions (*i.e.*, the Lévy solutions). The Rayleigh–Ritz and Galerkin methods can also be used to determine approximate analytical solutions, but they too are limited to simple geometries because of the difficulty in constructing the approximation functions for complicated geometries. The use of numerical methods facilitates the solution of these equations for problems of practical importance. Among the numerical methods available for the solution of differential equations defined over arbitrary domains, the finite element method is the most effective one [9,14,15].

There are several types of finite element models developed for plate theories. These can be grouped into three major categories: (i) displacement models, (ii) mixed and hybrid models, and (iii) equilibrium models. The displacement finite element models of plate theories are based on the principles of virtual displacements, where all governing equations are expressed in terms of the displacements [see, for example, Eqs. (2.4–20)–(2.4–27)]. The mixed and hybrid finite element models are based on modified or mixed variational statements of the plate theories, in which both displacements and stresses are independently approximated. The equilibrium models are based on the principle of virtual forces. Among the three types of models, the displacement finite element models are most natural and commonly used in commercial finite element programs.

The objective of this chapter is two-fold: first, to present the analytical solutions (the Navier as well as the Lévy solutions) for bending, buckling, and natural vibrations of rectangular plates; second, to develop the displacement finite element models of the classical laminated plate theory (CLPT) and the first-order shear deformation plate theory (FSDT). The Navier solutions are presented here only for antisymmetric cross-ply laminates using the classical plate theory. The Lévy solutions as well as the finite element models are presented for both CLPT and FSDT.

5.2 THE NAVIER SOLUTIONS

5.2–1 Boundary Conditions

Consider the following boundary conditions (SS1) on the generalized displacements and stress resultants of the classical laminate theory (see Figure 5.2–1):

$$
\begin{gathered}
u_0(x,0,t)=0\ ,\ \frac{\partial w_0}{\partial x}(x,0,t)=0\ ,\ u_0(x,b,t)=0\ ,\ \frac{\partial w_0}{\partial x}(x,b,t)=0 \\
v_0(0,y,t)=0\ ,\ \frac{\partial w_0}{\partial y}(0,y,t)=0\ ,\ v_0(a,y,t)=0\ ,\ \frac{\partial w_0}{\partial y}(a,y,t)=0 \\
w_0(x,0,t)=0\ ,\ w_0(x,b,t)=0\ ,\ w_0(0,y,t)=0\ ,\ w_0(a,y,t)=0
\end{gathered}
\tag{5.2 - 1a}
$$

$$
\begin{gathered}
N_{xx}(0,y,t)=0\ ,\ N_{xx}(a,y,t)=0\ ,\ N_{yy}(x,0,t)=0\ ,\ N_{yy}(x,b,t)=0 \\
M_{xx}(0,y,t)=0\ ,\ M_{xx}(a,y,t)=0\ ,\ M_{yy}(x,0,t)=0\ ,\ M_{yy}(x,b,t)=0
\end{gathered}
\tag{5.2 - 1b}
$$

The Navier solutions for rectangular laminated plates with the SS1 boundary conditions exist only when the laminate stacking sequences are such that

$$
A_{16}=0,\ A_{26}=0,\ B_{16}=0,\ B_{26}=0,\ D_{16}=0,\ D_{26}=0 \tag{5.2 - 2}
$$

Thus plates with a single generally orthotropic layer, symmetrically laminated plates with multiple specially orthotropic layers, and antisymmetric cross–ply laminated plates admit the Navier solutions for the SS1 boundary conditions.

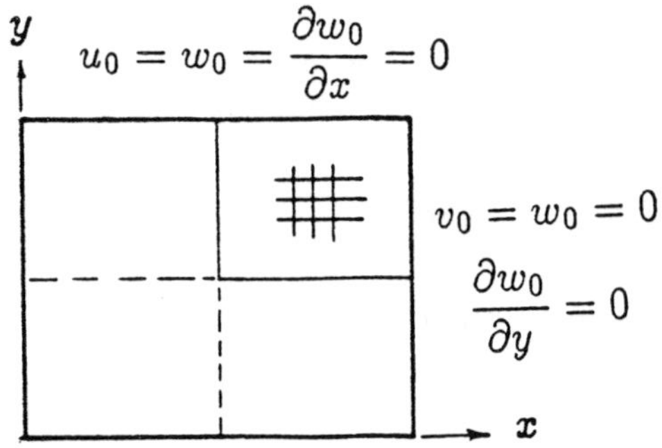

At a line of symmetry with normal **n**: $u_n = 0$
At an edge with normal **n**: $u_s = 0$ and w_0=0

Figure 5.2–1: Simply supported boundary conditions (SS–1) used for antisymmetric cross–ply laminates.

5.2–2 General Solution

The boundary conditions in (1) are satisfied by assuming the following form of the displacements

$$u_0(x, y, t) = \sum_{n=1}^{\infty}\sum_{m=1}^{\infty} U_{mn}(t) \cos\alpha x \sin\beta y \qquad (5.2-3a)$$

$$v_0(x, y, t) = \sum_{n=1}^{\infty}\sum_{m=1}^{\infty} V_{mn}(t) \sin\alpha x \cos\beta y \qquad (5.2-3b)$$

$$w_0(x, y, t) = \sum_{n=1}^{\infty}\sum_{m=1}^{\infty} W_{mn}(t) \sin\alpha x \sin\beta y \qquad (5.2-3c)$$

and the transverse distributed load is expanded in double trigonometric series as

$$q(x,y,t) = \sum_{n=1}^{\infty}\sum_{m=1}^{\infty} q_{mn}(t) f_{mn}^{1}(x,y) \tag{5.2-4}$$

where $\alpha = m\pi/a$ and $\beta = n\pi/b$. In a strict sense, α and β should have the subscripts m and n, respectively; *i.e.*, α_m and β_n. In the interest of brevity, they are omitted.

Substituting the expansions (3) and (4) into the governing equations (4.2–21)–(4.2–23) of the classical laminated plate theory (when $\hat{N}_{xx} = \hat{N}_{yy} = \hat{N}_{xy} = 0$), we obtain

$$\begin{bmatrix} \hat{c}_{11} & \hat{c}_{12} & \hat{c}_{13} \\ \hat{c}_{21} & \hat{c}_{22} & \hat{c}_{23} \\ \hat{c}_{31} & \hat{c}_{32} & \hat{c}_{33} \end{bmatrix} \begin{Bmatrix} U_{mn} \\ V_{mn} \\ W_{mn} \end{Bmatrix} + \begin{bmatrix} \hat{m}_{11} & 0 & \hat{m}_{13} \\ 0 & \hat{m}_{22} & \hat{m}_{23} \\ \hat{m}_{31} & \hat{m}_{32} & \hat{m}_{33} \end{bmatrix} \begin{Bmatrix} \ddot{U}_{mn} \\ \ddot{V}_{mn} \\ \ddot{W}_{mn} \end{Bmatrix} = \begin{Bmatrix} 0 \\ 0 \\ q_{mn} \end{Bmatrix} \tag{5.2-5}$$

where $c_{ij} = c_{ji}$ and $m_{ij} = m_{ji}$ (symmetric), and

$$\begin{aligned} \hat{c}_{11} &= (A_{11}\alpha^2 + A_{66}\beta^2) \\ \hat{c}_{12} &= (A_{12} + A_{66})\alpha\beta \\ \hat{c}_{13} &= -B_{11}\alpha^3 - (B_{12} + 2B_{66})\alpha\beta^2 \\ \hat{c}_{22} &= (A_{66}\alpha^2 + A_{22}\beta^2) \\ \hat{c}_{23} &= -(B_{12} + 2B_{66})\alpha^2\beta - B_{22}\beta^3 \\ \hat{c}_{33} &= D_{11}\alpha^4 + 2(D_{12} + 2D_{66})\alpha^2\beta^2 + D_{22}\beta^4 \\ \hat{m}_{11} &= I_0 \ , \ \hat{m}_{13} = -\alpha I_1 \ , \ \hat{m}_{22} = I_0 \\ \hat{m}_{23} &= -\beta I_1, \ \hat{m}_{33} = \Big(I_0 + I_2(\alpha^2 + \beta^2)\Big) \end{aligned} \tag{5.2-6}$$

Equations (6) and (7) provide three second-order differential equations in time among the three variables U_{mn}, V_{mn}, and W_{mn} for any fixed values of m and n. For transient (*i.e.*, time-dependent) response, the ordinary differential equations in time should be solved either exactly or approximately.

5.2–3 Bending Analysis

The steady–state (or equilibrium) solution can be obtained by solving the algebraic equations resulting from Eqs. (6) by setting the time derivative terms to zero:

$$\begin{bmatrix} \hat{c}_{11} & \hat{c}_{12} & \hat{c}_{13} \\ \hat{c}_{21} & \hat{c}_{22} & \hat{c}_{23} \\ \hat{c}_{31} & \hat{c}_{32} & \hat{c}_{33} \end{bmatrix} \begin{Bmatrix} U_{mn} \\ V_{mn} \\ W_{mn} \end{Bmatrix} = \begin{Bmatrix} 0 \\ 0 \\ q_{mn} \end{Bmatrix} \qquad (5.2-7)$$

Solution of Eq. (7) for each $m, n = 1, 2, \ldots$ gives (U_{mn}, V_{mn}, W_{mn}), which can then be used to compute the solution (u_0, v_0, w_0) from Eq. (3). The Navier solution for sinusoidally distributed transverse load or temperature field is a one–term solution ($m = n = 1$), and, therefore, it is a closed–form solution. For a general distribution of mechanical loads, the Navier solution is a series solution, which is evaluated for a sufficient number of terms in the series.

The stresses in each layer can be calculated from

$$\begin{Bmatrix} \sigma_{xx} \\ \sigma_{yy} \\ \sigma_{xy} \end{Bmatrix}^{(k)} = \sum_{m=1}^{\infty}\sum_{n=1}^{\infty} \begin{bmatrix} \bar{Q}_{11} & \bar{Q}_{12} & 0 \\ \bar{Q}_{12} & \bar{Q}_{22} & 0 \\ 0 & 0 & \bar{Q}_{66} \end{bmatrix}^{(k)} \begin{Bmatrix} S^{xx}_{mn} \sin\alpha x \sin\beta y \\ S^{yy}_{mn} \sin\alpha x \sin\beta y \\ S^{xy}_{mn} \cos\alpha x \cos\beta y \end{Bmatrix} \qquad (5.2-8)$$

where

$$\begin{aligned} S^{xx}_{mn} &= \left[-\alpha U_{mn} + z\left(\alpha^2 W_{mn}\right)\right] \\ S^{yy}_{mn} &= \left[-\beta V_{mn} + z\left(\beta^2 W_{mn}\right)\right] \\ S^{xy}_{mn} &= \left[\beta U_{mn} + \alpha V_{mn} - 2z\left(\alpha\beta W_{mn}\right)\right] \end{aligned} \qquad (5.2-9)$$

The maximum values of σ_{xx} and σ_{yy} occur at $(x, y) = (a/2, b/2)$, and the maximum of σ_{xy} occurs at $(x, y) = (a, b)$. The maxima with respect to the thickness coordinate vary with the lamination scheme.

The load coefficients Q_{mn} for various types of loading [see Eq. (11b)] are listed in Table 5.2–1. In particular, for uniformly distributed load, $q(x,y) = q_0$, a constant, we have

$$Q_{mn} = \frac{16q_0}{\pi^2 mn} \quad \text{for } m, n, \text{ odd} \tag{5.2 – 10}$$

For a transverse point load Q_0 located at (x_0, y_0), the load coefficients are given by $[q(x,y) = Q_0\delta(x - x_0, y - y_0)]$

$$Q_{mn} = \frac{4Q_0}{ab} \sin\frac{m\pi x_0}{a} \sin\frac{n\pi y_0}{b} \tag{5.2 – 11}$$

5.2–4 Buckling Analysis

When a rectangular laminated plate is subjected to an inplane compressive load of sufficiently small magnitude, the equilibrium of the plate is stable. As the magnitude of the inplane force increases, the stable state of the plate is disturbed (*i.e.*, becomes unstable), and the plate is said to have buckled. The magnitude of the compressive load at which the plate becomes unstable is termed the *critical buckling load.* If the load is further increased, it results in large transverse deflection and leads to failure of the plate. The Navier solution for buckling can be obtained by modifying the governing equations by accounting for the inplane compressive load.

For laminated plates subjected to compressive inplane forces $\hat{N}_{xx} = -N^0_{xx}$ and $\hat{N}_{yy} = -N^0_{yy}$, with

$$\hat{N}_{xx} = -N_0\ , \quad \hat{N}_{yy} = -kN_0\ , \quad k = \frac{\hat{N}_{yy}}{\hat{N}_{xx}} \tag{5.2 – 12}$$

$$\begin{bmatrix} \hat{c}_{11} & \hat{c}_{12} & \hat{c}_{13} \\ \hat{c}_{12} & \hat{c}_{22} & \hat{c}_{23} \\ \hat{c}_{13} & \hat{c}_{23} & \hat{c}_{33} - N_0(\alpha^2 + k\beta^2) \end{bmatrix} \begin{Bmatrix} U_{mn} \\ V_{mn} \\ W_{mn} \end{Bmatrix} = \begin{Bmatrix} 0 \\ 0 \\ 0 \end{Bmatrix} \tag{5.2 – 13}$$

Table 5.2–1 Coefficients in the double trigonometric series expansion of loads in the Navier method.

$$Q_{mn} = \frac{4}{ab}\int_0^a \int_0^b q(x,y)\,\sin\frac{m\pi x}{a}\,\sin\frac{n\pi y}{b}\,dxdy$$

Load $q(x,y)$	Coefficients Q_{mn}
Uniform load, $q = -q_0$ 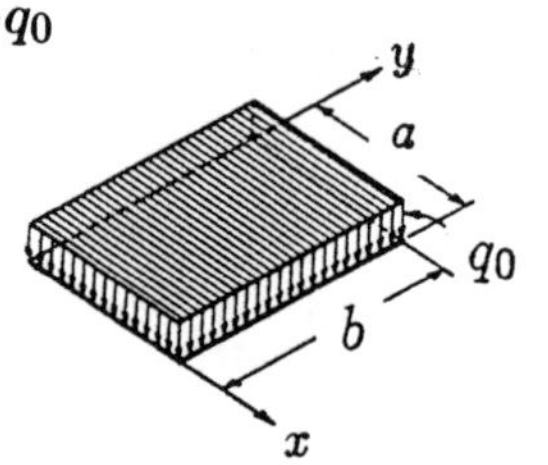	$Q_{mn} = -\frac{16q_0}{\pi^2 mn}$ $(m, n = 1, 3, 5, \cdots)$
Hydrostatic load, $q(x,y) = -q_0\frac{x}{a}$ 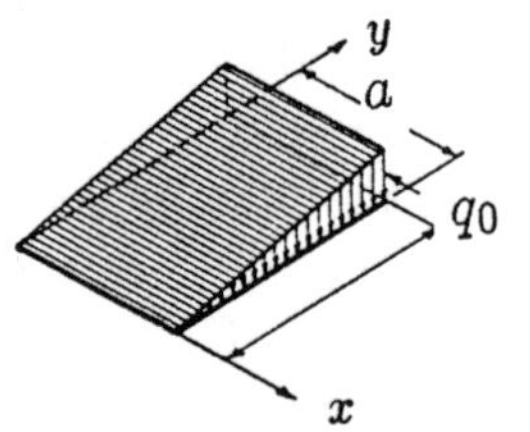	$Q_{mn} = -\frac{8q_0 \cos m\pi}{\pi^2 mn}$ $(m, n = 1, 3, 5, \cdots)$
Point load, $q(x,y) = -Q_0$ at (x_0, y_0) 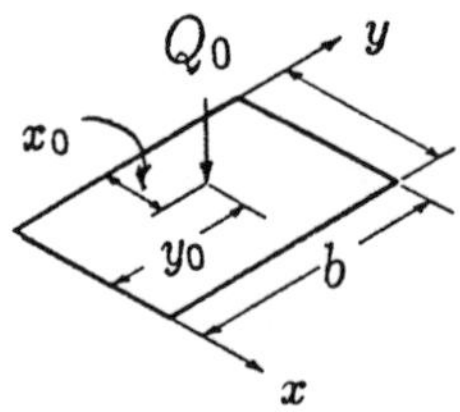	$Q_{mn} = -\frac{4Q_0}{ab} f(x_0, y_0)$ $f = \sin\frac{m\pi x_0}{a}\sin\frac{n\pi y_0}{b}$ $(m, n = 1, 2, 3, \cdots)$
Line load, $q(x,y) = -q_0$ at $x = x_0$ 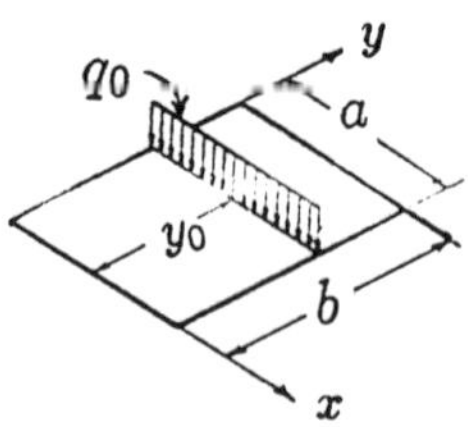	$Q_{mn} = -\frac{4q_0}{\pi an}\sin\frac{m\pi x_0}{a}$ $(m, n = 1, 2, 3, \cdots)$

where $\hat{c}_{ij}$ are the coefficients defined in Eq. (6b). For nontrivial solution, $U_{mn} \neq 0$, $V_{mn} \neq 0$, and $W_{mn} \neq 0$, the determinant of the coefficient matrix in (13) should be zero. We obtain

$$N_0 = \frac{1}{\alpha^2 + k\beta^2}\left(\hat{c}_{33} - \frac{\hat{c}_{13}\hat{c}_{22} - \hat{c}_{23}\hat{c}_{12}}{\hat{c}_{11}\hat{c}_{22} - \hat{c}_{12}\hat{c}_{12}}\hat{c}_{13} - \frac{\hat{c}_{11}\hat{c}_{23} - \hat{c}_{12}\hat{c}_{13}}{\hat{c}_{11}\hat{c}_{22} - \hat{c}_{12}\hat{c}_{12}}\hat{c}_{23}\right) \tag{5.2 - 14}$$

Clearly, for each pair of m and n, there is a unique value of N_0. The critical buckling load is the smallest of all $N_0 = N_0(m, n)$. For a given laminate (*i.e.*, for a given set of material properties, stacking sequence, and geometry), this value is dictated by a particular combination of the values of m and n (see Reddy [4] for additional details).

For specially orthotropic plates, neither shear nor twist coupling nor bending–extension coupling exists ($\hat{c}_{13} = \hat{c}_{23} = 0$), and therefore U_{mn} and V_{mn} are zero prior to onset of buckling. Therefore, we have

$$N_0 = \frac{\hat{c}_{33}}{(\alpha^2 + k\beta^2)}$$

where

$$\hat{c}_{33} = D_{11}\alpha^4 + 2(D_{12} + 2D_{66})\alpha^2\beta^2 + D_{22}\beta^4 \ , \ \alpha = \frac{m\pi}{a} \ , \quad \beta = \frac{m\pi}{b}$$

For a square laminate subjected to the same magnitude of compressive load on both edges (*i.e.*, biaxial compression with $k = 1$), Eq. (14) yields

$$N_0(m, n) = \frac{\pi^2}{a^2}\frac{[D_{11}m^4 + 2(D_{12} + 2D_{66})m^2n^2 + D_{22}n^4]}{m^2 + n^2}$$

For modulus ratios of $D_{11}/D_{22} = 10$ and $(D_{12} + 2D_{22})/D_{22} = 1$, for example, the critical buckling load becomes (see Reddy [4])

$$N_{cr} = 6.5\frac{\pi^2 D_{22}}{a^2}$$

5.2–5 Vibration Analysis

For free vibration, we set all applied loads and the inplane forces to zero and assume a periodic solution of the form

$$\begin{aligned} U_{mn}(t) &= U_{mn}^0 e^{i\omega t} \\ V_{mn}(t) &= V_{mn}^0 e^{i\omega t} \\ W_{mn}(t) &= W_{mn}^0 e^{i\omega t} \end{aligned} \tag{5.2 – 15}$$

where $i = \sqrt{-1}$, and ω is the frequency of natural vibration. Then Eq. (6a) reduces to

$$\left(\begin{bmatrix} \hat{c}_{11} & \hat{c}_{12} & \hat{c}_{13} \\ \hat{c}_{12} & \hat{c}_{22} & \hat{c}_{23} \\ \hat{c}_{13} & \hat{c}_{23} & \hat{c}_{33} \end{bmatrix} - \omega^2 \begin{bmatrix} \hat{m}_{11} & 0 & 0 \\ 0 & \hat{m}_{22} & 0 \\ 0 & 0 & \hat{m}_{33} \end{bmatrix} \right) \begin{Bmatrix} U_{mn}^0 \\ V_{mn}^0 \\ W_{mn}^0 \end{Bmatrix} = \begin{Bmatrix} 0 \\ 0 \\ 0 \end{Bmatrix} \tag{5.2 – 16}$$

where $\hat{c}_{ij}$ and $\hat{m}_{ij}$ are defined in Eq. (6b). For nontrivial solution, $U_{mn}^0 \neq 0$, $V_{mn}^0 \neq 0$, $W_{mn}^0 \neq 0$, the determinant of the coefficient matrix in (16) should be zero, which yields the characteristic polynomial

$$-p\lambda^3 + q\lambda^2 - r\lambda + s = 0 \tag{5.2 – 17a}$$

where $\lambda = \omega^2$, and

$$p = \begin{vmatrix} \hat{m}_{11} & 0 & 0 \\ 0 & \hat{m}_{22} & 0 \\ 0 & 0 & \hat{m}_{33} \end{vmatrix}$$

$$q = \begin{vmatrix} \hat{c}_{11} & 0 & 0 \\ \hat{c}_{12} & \hat{m}_{22} & 0 \\ \hat{c}_{13} & 0 & \hat{m}_{33} \end{vmatrix} + \begin{vmatrix} \hat{m}_{11} & \hat{c}_{12} & 0 \\ \hat{0} & \hat{c}_{22} & 0 \\ 0 & \hat{c}_{23} & \hat{m}_{33} \end{vmatrix} + \begin{vmatrix} \hat{m}_{11} & 0 & \hat{c}_{13} \\ \hat{0} & \hat{m}_{22} & \hat{c}_{23} \\ 0 & 0 & \hat{c}_{33} \end{vmatrix}$$

$$r = \begin{vmatrix} \hat{c}_{11} & \hat{c}_{12} & 0 \\ \hat{c}_{12} & \hat{c}_{22} & 0 \\ \hat{c}_{13} & \hat{c}_{23} & \hat{m}_{33} \end{vmatrix} + \begin{vmatrix} \hat{c}_{11} & \hat{0} & \hat{c}_{13} \\ \hat{c}_{12} & \hat{m}_{22} & \hat{c}_{23} \\ \hat{c}_{13} & 0 & \hat{c}_{33} \end{vmatrix} + \begin{vmatrix} \hat{m}_{11} & \hat{c}_{12} & \hat{c}_{13} \\ \hat{0} & \hat{c}_{22} & \hat{c}_{23} \\ 0 & \hat{c}_{23} & \hat{c}_{33} \end{vmatrix}$$

$$s = \begin{vmatrix} \hat{c}_{11} & \hat{c}_{12} & \hat{c}_{13} \\ \hat{c}_{12} & \hat{c}_{22} & \hat{c}_{23} \\ \hat{c}_{13} & \hat{c}_{23} & \hat{c}_{33} \end{vmatrix} \qquad (5.2-17b)$$

The real positive roots of this cubic equation give the square of the natural frequency ω_{mn} associated with mode (m, n). The smallest of the frequencies is called the *fundamental frequency.* In general, ω_{11} is not the fundamental frequency; the smallest frequency might occur for values other than $m = n = 1$.

For specially orthtropic laminated plates, the only nonzero stiffnesses are $A_{11}, A_{12}, A_{22}, A_{66}, D_{11}, D_{12}, D_{22}$, and D_{66}. Consequently, $\hat{c}_{13}$ and $\hat{c}_{23}$ are zero. Hence, Eq. (16) reduces to

$$\begin{vmatrix} \hat{c}_{11} - \omega^2 \hat{m}_{11} & \hat{c}_{12} & 0 \\ \hat{c}_{12} & \hat{c}_{22} - \omega^2 \hat{m}_{22} & 0 \\ 0 & 0 & \hat{c}_{33} - \omega^2 \hat{m}_{33} \end{vmatrix} = 0 \qquad (5.2-18a)$$

which yields the characteristic polynomial

$$\left(\hat{c}_{33} - \omega^2 \hat{m}_{33}\right) \left[\left(\hat{c}_{11} - \omega^2 \hat{m}_{11}\right) \left(\hat{c}_{22} - \omega^2 \hat{m}_{22}\right) - \hat{c}_{12}\hat{c}_{12}\right] = 0 \qquad (5.2-18b)$$

The cubic polynomial has three roots:

$$(\omega_{mn}^2)_1 = \frac{\hat{c}_{33}}{\hat{m}_{33}} \qquad (5.2-19a)$$

$$(\omega_{mn}^2)_2 = \frac{1}{2} \left[\hat{c}_{11} + \hat{c}_{22} - \sqrt{(\hat{c}_{11} - \hat{c}_{22})^2 + 4\hat{c}_{12}\hat{c}_{12}}\right] \qquad (5.2-19b)$$

$$(\omega_{mn}^2)_3 = \frac{1}{2} \left[\hat{c}_{11} + \hat{c}_{22} + \sqrt{(\hat{c}_{11} - \hat{c}_{22})^2 + 4\hat{c}_{12}\hat{c}_{12}}\right] \qquad (5.2-19c)$$

The first frequency corresponds to the flexural vibration, and the last two correspond to the inplane vibration. It can be easily verified that the frequencies of inplane vibration are much higher than those of the flexural vibration.

The frequencies of flexural vibration can be written in explicit form as

$$\omega_{mn} = \frac{\pi^4}{\tilde{I}_0 b^4}\left[D_{11}m^4\left(\frac{b}{a}\right)^4 + 2(D_{12}+2D_{66})m^2n^2\left(\frac{b}{a}\right)^2 + D_{22}n^4\right] \tag{5.2 - 20a}$$

where

$$\tilde{I}_0 = I_0 + I_2\left[\left(\frac{m\pi}{a}\right)^2 + \left(\frac{n\pi}{b}\right)^2\right]$$

$$I_0 = \sum_{k=1}^{L} \rho_0^{(k)}\left(z_{k+1} - z_k\right)\ ,\quad I_2 = \frac{1}{3}\sum_{k=1}^{L}\rho_0^{(k)}\left(z_{k+1}^3 - z_k^3\right) \tag{5.2 - 20b}$$

and L denotes the total number of layers in the laminate. For different values of m and n there corresponds a unique frequency ω_{mn} and corresponding mode shape

$$w_0(x,y) = \sum_{m=1}^{\infty}\sum_{m=1}^{\infty} W_{mn}^0 \sin\frac{m\pi x}{a}\ \sin\frac{n\pi y}{b} \tag{5.2 - 21}$$

where W_{mn}^0 is the amplitude of the vibration mode (m,n), which is seldom needed because one is interested in the frequency and mode shape only.

It should be noted that the fundamental mode (*i.e.*, the smallest frequency of vibration) is not always associated with $m = n = 1$. It depends on the material properties, lamination scheme, geometry, and boundary conditions. It is not easy, in general, to determine analytically the values of m and n for which the frequency of vibration is the minimum.

5.3 THE LÉVY SOLUTIONS

5.3–1 Introduction

The Lévy method is one in which the partial differential equations of laminated plates are reduced to a system of ordinary differential equations by applying the Navier method with respect to one of the coordinates (*i.e.*, the boundary conditions and governing equations are exactly satisfied with respect to one coordinate). The resulting ordinary differential equations may be solved using the *state–space approach.* In this approach an equation of n–th order is reduced to a system of n first–order equations, which are then cast in matrix form as a first–order equation. The first–order matrix equation is then solved in the same way as a single first–order equation, except that the eigenvalues and eigenvectors associated with the matrix operator are required.

5.3–2 Operator Form of the Governing Equations

For the static case, the governing equations appropriate for the antisymmetric cross-ply laminate construction are given by

$$[L]\{\Delta\} = \{F\} \qquad (5.3-1)$$

where

$$\{\Delta\} = \begin{Bmatrix} u_0 \\ v_0 \\ w_0 \end{Bmatrix} \quad , \quad \{F\} = \begin{Bmatrix} 0 \\ 0 \\ q \end{Bmatrix} \qquad (5.3-2)$$

for CLPT, and

$$\{\Delta\} = \begin{Bmatrix} u_0 \\ v_0 \\ w_0 \\ \phi_x \\ \phi_y \end{Bmatrix} \quad , \quad \{F\} = \begin{Bmatrix} 0 \\ 0 \\ q \\ 0 \\ 0 \end{Bmatrix} \qquad (5.3-3)$$

for FSDT, and the matrix coefficients L_{ij} for the two theories are listed below.

Coefficients L_{ij} for CLPT:

$$\begin{aligned}
L_{11} &= A_{11}d_1^2 + A_{66}d_2^2 \\
L_{12} &= (A_{12} + A_{66})d_1 d_2 \\
L_{13} &= -B_{11}d_1^3 - (B_{12} + 2B_{66})d_1 d_2^2 \\
L_{22} &= A_{66}d_1^2 + A_{22}d_2^2 \\
L_{23} &= -B_{22}d_2^3 - (B_{12} + 2B_{66})d_1^2 d_2 \\
L_{33} &= D_{11}d_1^4 + 2(D_{12} + 2D_{66})d_1^2 d_2^2 + D_{22}d_2^4 \qquad (5.3-4)
\end{aligned}$$

Coefficients L_{ij} for FSDT:

$$\begin{aligned}
L_{11} &= A_{11}d_1^2 + A_{66}d_2^2 \\
L_{12} &= (A_{12} + A_{66})d_1 d_2 \\
L_{13} &= 0 \\
L_{14} &= B_{11}d_1^2 + B_{66}d_2^2 \\
L_{15} &= (B_{12} + B_{66})d_1 d_2 \\
L_{22} &= A_{66}d_1^2 + A_{22}d_2^2 \\
L_{23} &= 0 \\
L_{24} &= L_{15} \\
L_{25} &= B_{22}d_2^2 + B_{66}d_1^2 \\
L_{33} &= -K_{55}^2 A_{55}d_1^2 - K_{44}^2 A_{44}d_2^2 \\
L_{34} &= -K_{55}^2 A_{55}d_1 \\
L_{35} &= -K_{44}^2 A_{44}d_2 \\
L_{44} &= -K_{55}^2 A_{55} + D_{11}d_1^2 + D_{66}d_2^2 \\
L_{45} &= (D_{12} + D_{66})d_1 d_2 \\
L_{55} &= -K_{44}^2 A_{44} + D_{22}d_2^2 + D_{66}d_1^2 \qquad (5.3-5)
\end{aligned}$$

where

$$d_\alpha^i = \frac{\partial^i}{\partial x_\alpha^i}, \quad x_1 = x \text{ and } x_2 = y \qquad (5.3-6)$$

5.3–3 Solution Procedure

The Lévy type solution can be developed for rectangular laminates with the following boundary conditions: The edges $y = 0, b$ are simply supported, while the remaining ones ($x = \pm a/2$) may have arbitrary combinations of free, clamped, and simply supported edge conditions (see Figure 5.3–1). The generalized displacements may be expressed as products of undetermined functions and known trigonometric functions so as to satisfy identically the simply supported boundary conditions at $y = 0, b$:

$$u_0 = w_0 = N_{yy} = M_{yy} = 0 \text{ for CLPT} \qquad (5.3-7)$$

$$u_0 = w_0 = \phi_x = N_{yy} = M_{yy} = 0 \text{ for FSDT} \qquad (5.3-8)$$

For this case one may be able to develop the Lévy type solutions for the antisymmetric cross-ply lamination scheme (see [1]).

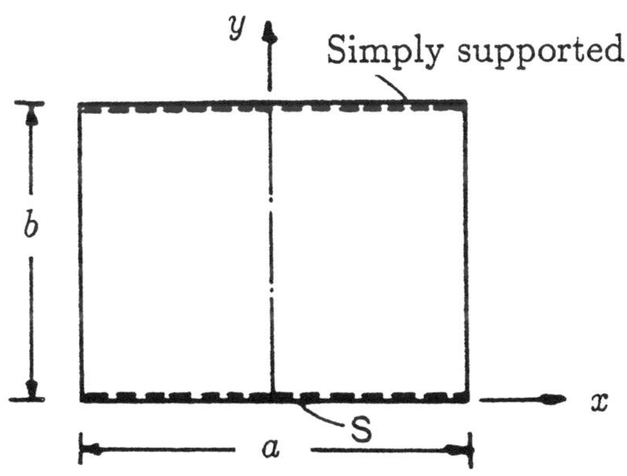

Figure 5.3–1: Coordinate system used in the Lévy solution procedure.

A sinusoidal distribution of the transverse load is considered, which for the present case takes the form

$$q(x,y) = q_0 \cos\frac{\pi x}{a} \sin\frac{\pi y}{b} \tag{5.3 – 9}$$

The displacement functions are represented as

$$\begin{Bmatrix} u_0(x,y) \\ v_0(x,y) \\ w_0(x,y) \\ \phi_x(x,y) \\ \phi_y(x,y) \end{Bmatrix} = \sum_{m=1}^{\infty} \begin{Bmatrix} U_m(x)\sin\beta y \\ V_m(x)\cos\beta y \\ W_m(x)\sin\beta y \\ X_m(x)\sin\beta y \\ Y_m(x)\cos\beta y \end{Bmatrix} \tag{5.3 – 10}$$

where $\beta \equiv m\pi/b$.

Substitution of (10) into Eqs. (1) results in three differential equations for CLPT and five differential equations for FSDT. We have

CLPT:

$$\begin{aligned} U''_m &= C_1 U_m + C_2 V'_m + C_3 W'_m + C_4 W'''_m \\ V''_m &= C_5 U'_m + C_6 V_m + C_7 W_m + C_8 W''_m \\ W''''_m &= C_9 U'_m + C_{10} V_m + C_{11} W_m + C_{12} W''_m - f/e_{18} \end{aligned} \tag{5.3 – 11a}$$

$$\begin{aligned} C_1 &= -e_2/e_1, \quad C_2 = -e_3/e_1, \quad C_3 = -e_5/e_1 \\ C_4 &= -e_4/e_1, \quad C_5 = -e_6/e_7, \quad C_6 = -e_8/e_7 \\ C_7 &= -e_{10}/e_7, \quad C_8 = -e_9/e_7, \quad C_9 = -e_{21}/e_{18} \\ C_{10} &= -e_{22}/e_{18}, \quad C_{11} = -e_{20}/e_{18}, \quad C_{12} = -e_{19}/e_{18} \end{aligned} \tag{5.3 – 11b}$$

$$\begin{aligned} e_1 &= A_{11}, \quad e_2 = -\beta^2 A_{66}, \quad e_3 = -\beta(A_{12} + A_{66}), \quad e_4 = -B_{11} \\ e_5 &= \beta^2(B_{12} + 2B_{66}), \quad e_6 = -e_3, \quad e_7 = A_{66}, \quad e_8 = -\beta^2 A_{22} \\ e_9 &= \beta(B_{12} + 2B_{66}), \quad e_{10} = \beta^3 B_{22}, \quad e_{11} = D_{11} \\ e_{12} &= -2\beta^2(D_{12} + 2D_{66}), \quad e_{13} = \beta^4 D_{22} \\ e_{14} &= e_4, \; e_{15} = e_5, \; e_{16} = -e_9, \; e_{17} = -e_{10}, \; e_{18} = e_{11} - e_4 e_{14}/e_1 \\ e_{19} &= e_{12} - e_5 e_{14}/e_1 - e_9 e_{16}/e_7 + e_3 e_9 e_{14}/(e_1 e_7) \\ e_{20} &= e_{13} - e_{10} e_{16}/e_7 + e_3 e_{10} e_{14}/(e_1 e_7) \\ e_{21} &= e_{15} - e_2 e_{14}/e_1 - e_6 e - 16/e_7 + e_3 e_6 e_{14}/(e_1 e_7) \\ e_{22} &= e_{17} - e_8 e_{16}/e_7 + e_3 e_8 e_{14}/(e_1 e_7) \end{aligned} \tag{5.3 – 11c}$$

FSDT:

$$
\begin{aligned}
U''_m &= C_1 U_m + C_2 V'_m + C_3 W'_m + C_4 X_m + C_5 Y'_m \\
V''_m &= C_6 U'_m + C_7 V_m + C_8 W_m + C_9 X'_m + C_{10} Y_m \\
W''_m &= C_{11} W_m + C_{12} X'_m + C_{13} Y_m - f/e_{13} \\
X''_m &= C_{14} U_m + C_{15} V'_m + C_{16} W'_m + C_{17} X_m + C_{18} Y'_m \\
Y''_m &= C_{19} U'_m + C_{20} V_m + C_{21} W_m + C_{22} X'_m + C_{23} Y_m \qquad (5.3-12a)
\end{aligned}
$$

$$
\begin{aligned}
C_1 &= (e_3 e_{21} - e_5 e_{19})/e_0, \quad C_2 = (e_3 e_{18} - e_2 e_{19})/e_0, \\
C_3 &= e_3 e_{23}/e_0, \quad C_4 = (e_3 e_{22} - e_6 e_{19})/e_0 \\
C_5 &= (e_3 e_{20} - e_4 e_{19})/e_0, \quad C_6 = (e_8 e_{27} - e_{10} e_{25})/C_0 \\
C_7 &= (e_{11} e_{27} - e_{10} e_{28})/C_0, \quad C_8 = -e_{10} e_{30}/C_0, \\
C_9 &= (e_9 e_{27} - e_{10} e_{26})/C_0, \quad C_{10} = (e_{12} e_{27} - e_{10} e_{29})/C_0, \\
C_{11} &= -e_{15}/e_{13}, \\
C_{12} &= -e_{14}/e_{13}, \quad C_{13} = -e_{16}/e_{13}, \quad C_{14} = (e_5 e_{17} - e_1 e_{21})/e_0, \\
C_{15} &= (e_2 e_{17} - e_1 e_{18})/e_0, \quad C_{16} = -e_1 e_{23}/e_0, \\
C_{17} &= (e_6 e_{17} - e_1 e_{22})/e_0, \quad C_{18} = (e_4 e_{17} - e_1 e_{20})/e_0, \\
C_{19} &= (e_7 e_{25} - e_8 e_{24})/C_0, \quad C_{20} = (e_7 e_{28} - e_{11} e_{24})/C_0, \\
C_{21} &= e_7 e_{30}/C_0, \quad C_{22} = (e_7 e_{26} - e_9 e_{24})/C_0, \\
C_{23} &= (e_7 e_{29} - e_{12} e_{24})/C_0, \quad e_0 = e_1 e_{19} - e_3 e_{17}, \\
C_0 &= e_{10} e_{24} - e_7 e_{27} \qquad (5.3-12b)
\end{aligned}
$$

$$
\begin{aligned}
e_1 &= A_{11}, \quad e_2 = -\beta(A_{12} + A_{66}), \quad e_3 = B_{11}, \quad e_4 = -\beta(B_{12} + B_{66}) \\
e_5 &= -\beta^2 A_{66}, \quad e_6 = -\beta^2 B_{66}, \quad e_7 = A_{66}, \quad e_8 = -e_2 \\
e_9 &= -e_4, \quad e_{10} = B_{66}, \quad e_{11} = -\beta^2 A_{22}, \quad e_{12} = -\beta^2 B_{22} \\
e_{13} &= K_{55}^2 A_{55}, \quad e_{14} = K_{55}^2 A_{55}, \quad e_{15} = -\beta^2 K_{44}^2 A_{44} \\
e_{16} &= -\beta K_{44}^2 A_{44}, \quad e_{17} = e_3, \quad e_{18} = e_4, \quad e_{19} = D_{11} \\
e_{20} &= -\beta(D_{11} + D_{66}), \quad e_{21} = -\beta^2 B_{66}, \quad e_{22} = -\beta^2 D_{66} - K_{55}^2 A_{55} \\
e_{23} &= -e_{14}, \quad e_{24} = e_{10}, \quad e_{25} = -e_4, \quad e_{26} = -e_{20}, \quad e_{27} = D_{66} \\
e_{28} &= -\beta^2 B_{22}, \quad e_{29} = -\beta^2 D_{22} - K_{44}^2 A_{44}, \quad e_{30} = e_{16} \qquad (5.3-12c)
\end{aligned}
$$

where $(\ )'$ denotes the derivative with respect to x.

In order to reduce the system of equations (11) and (12) to the first–order system

$$\mathbf{Z}' = \mathbf{AZ} + \mathbf{r} \tag{5.3 – 13}$$

the components of the state vector $\mathbf{Z}(x)$, the operator $[A]$, and vector $\{r\}$ associated with each theory are defined below.

CLPT:

$$Z_1 = U_m; \quad Z_2 = U'_m; \quad Z_3 = V_m; \quad Z_4 = V'_m; \quad Z_5 = W_m; \quad Z_6 = W'_m;$$

$$Z_7 = W''_m; \quad Z_8 = W'''_m \tag{5.3 – 14a}$$

$$\{r\} = \{0, 0, 0, 0, 0, 0, 0, f/e_{18}\}^T \tag{5.3 – 14b}$$

$$[A] = \begin{bmatrix} 0 & 1 & 0 & 0 & 0 & 0 & 0 & 0 \\ C_1 & 0 & 0 & C_2 & 0 & C_3 & 0 & C_4 \\ 0 & 0 & 0 & 1 & 0 & 0 & 0 & 0 \\ 0 & C_5 & C_6 & 0 & C_7 & 0 & C_8 & 0 \\ 0 & C_5 & C_6 & 0 & C_7 & 0 & C_8 & 0 \\ 0 & 0 & 0 & 0 & 0 & 1 & 0 & 0 \\ 0 & 0 & 0 & 0 & 0 & 0 & 1 & 0 \\ 0 & 0 & 0 & 0 & 0 & 0 & 0 & 1 \\ 0 & C_9 & C_{10} & 0 & C_{11} & 0 & C_{12} & 0 \end{bmatrix} \tag{5.3 – 14c}$$

FSDT:

$$Z_1 = U_m, \quad Z_2 = U'_m, \quad Z_3 = V_m, \quad Z_4 = V'_m, \quad Z_5 = W_m, \quad Z_6 = W'_m$$

$$Z_7 = X_m, \quad Z_8 = X'_m, \quad Z_9 = Y_m, \quad Z_{10} = Y'_m, \tag{5.3 – 15a}$$

$$\{r\} = \{0, 0, 0, 0, 0, -f/e_{13}, 0, 0, 0, 0\}^T \tag{5.3 – 15b}$$

$$[A] = \begin{bmatrix} 0 & 1 & 0 & 0 & 0 & 0 & 0 & 0 & 0 & 0 \\ C_1 & 0 & 0 & C_2 & 0 & C_3 & C_4 & 0 & 0 & C_5 \\ 0 & 0 & 0 & 1 & 0 & 0 & 0 & 0 & 0 & 0 \\ 0 & C_6 & C_7 & 0 & C_8 & 0 & 0 & C_9 & C_{10} & 0 \\ 0 & 0 & 0 & 0 & 0 & 1 & 0 & 0 & 0 & 0 \\ 0 & 0 & 0 & 0 & C_{11} & 0 & 0 & C_{12} & C_{13} & 0 \\ 0 & 0 & 0 & 0 & 0 & 0 & 0 & 1 & 0 & 0 \\ C_{14} & 0 & 0 & C_{15} & 0 & C_{16} & C_{17} & 0 & 0 & C_{18} \\ 0 & 0 & 0 & 0 & 0 & 0 & 0 & 0 & 0 & 1 \\ 0 & C_{19} & C_{20} & 0 & C_{21} & 0 & 0 & C_{22} & C_{23} & 0 \end{bmatrix} \quad (5.3-15c)$$

where $f = q_0 \cos \frac{\pi x}{a}$.

The solution to Eq. (13) is given by (see [2,3])

$$\mathbf{Z} = e^{\mathbf{A}x}\{\mathbf{K} + \int e^{-\mathbf{A}\zeta}\mathbf{r}\, d\zeta\} \quad (5.3-16)$$

Here $\mathbf{K}$ is constant vector to be determined from the boundary conditions, and $e^{\mathbf{A}x}$ has the meaning

$$e^{\mathbf{A}x} = [S]\begin{bmatrix} e^{\lambda_1 x} & \mathbf{0} \\ \mathbf{0} & e^{\lambda_n x} \end{bmatrix}[S]^{-1} \quad (5.3-17)$$

where n denotes the order of the theory ($n = 8$ for CLPT and $n = 10$ for FSDT), and λ_i denote the distinct eigenvalues of $[A]$, while $[S]$ denotes the matrix of eigenvectors of $[A]$.

The boundary conditions for simply supported (S), clamped (C) and free (F) at the edges $x = \pm a/2$ for the two theories are given below.

CLPT:

$$S: \ v_0 = w_0 = N_{xx} = M_{xx} = 0 \quad (5.3-18a)$$

$$C: \ u_0 = v_0 = w_0 = \frac{\partial w_0}{\partial x} = 0 \quad (5.3-18b)$$

$$F: \ N_{xx} = M_{xx} = N_{xy} = 0\ , \ \frac{\partial M_{xx}}{\partial x} + \frac{\partial M_{xy}}{\partial y} = 0 \quad (5.3-18c)$$

FSDT:

$$S: \ v_0 = w_0 = \phi_y = N_{xx} = M_{xx} = 0 \qquad (5.3-19a)$$
$$C: \ u_0 = v_0 = w_0 = \phi_x = \phi_y = 0 \qquad (5.3-19b)$$
$$F: \ N_{xx} = M_{xx} = Q_x = N_{xy} = M_{xy} = 0 \qquad (5.3-19c)$$

Use of the boundary conditions in Eq. (16) results in a set of algebraic equations of the form

$$[B]\{K\} = \{R\} \qquad (5.3-20)$$

which can be solved for the vector $\{K\}$, and the solution is given by Eq. (16).

5.4 FINITE ELEMENT MODELS

5.4–1 Classical Laminated Plate Theory

The development of a finite element model follows two basic steps: (1) the construction of weighted–integral statements of the governing equations, and (2) approximation of the displacements, using ideas from the interpolation theory, and derivation of algebraic equations among the nodal displacements and forces. In the first step, the differential equations to be solved are recast in a weighted–integral form. This is necessary in order to derive a set of linearly independent algebraic relations among the nodal displacements and forces. The weighted–integral statement can be derived either directly from the governing equations (4.2–5)–(4.2–7) or from the principle of virtual displacements. Both approaches give exactly the same weighted–integral statement, called the *weak form.* In the second step, the dependent variables are approximated by appropriate finite element interpolations, and they are substituted into the weak form of the equations to obtain the algebraic relations among the nodal values of the generalized displacements and forces.

There are two basic types of interpolations that are used in finite element analyses. The Lagrange interpolation is one in which only the function is interpolated, whereas the Hermite interpolation is one in which the function and its derivatives are interpolated. The finite elements developed using the Lagrange type interpolation are called C^0 elements, and finite elements developed using Hermite type interpolation are called C^m elements, where $m > 0$ is the order of the derivatives included in the interpolation.

From the discussion of Section 4.2–3, it should be recalled that the primary variables of the classical plate theory are u_0, v_0, w_0, $\partial w_0/\partial x$, and $\partial w_0/\partial x$. In the finite element method, the primary variables are interpolated and they are made continuous across elements (see Reddy [4,5]). Hence, the displacements (u_0, v_0, w_0) are approximated over an element Ω^e of the domain Ω (*i.e.*, midsurface of the laminate) by interpolations of the form

$$u_0 = \sum_{j=1}^{n} u_j(t)\psi_j(x,y)$$
$$v_0 = \sum_{j=1}^{n} v_j(t)\psi_j(x,y)$$
$$w_0 = \sum_{j=1}^{m} \Delta_j(t)\phi_j(x,y) \qquad (5.4-1)$$

Here (u_j, v_j) denote the nodal values of (u_0, v_0) and Δ_j denote the nodal values of w_0 and its derivatives for any time t, and (ψ_j, ϕ_j) are associated interpolation functions. Since primary variables contain u_0 and v_0 and not their derivatives, it follows that ψ_i are the Lagrange interpolation functions (see Reddy [5,9]). On the other hand, since w_0 and its first derivatives are primary variables, ϕ_i are the Hermite interpolation functions.

For linear Lagrange interpolation of (u_0, v_0) and Hermite cubic interpolation of w_0 using four–node rectangular elements, we have $n = 4$ and $m = 16$, and the total number of nodal degrees of freedom per element is 24. The four nodal values associated with w_0 are

$$w_0, \ \frac{\partial w_0}{\partial x}, \ \frac{\partial w_0}{\partial y}, \ \text{and} \ \frac{\partial^2 w_0}{\partial x \partial y} \qquad (5.4-2)$$

The Lagrange interpolation functions ψ_i and Hermite interpolation functions ϕ_i are presented in Table 5.4–1 (taken from Table 7.9 of Reddy [9]; also see Reddy [10]).

Table 5.4–1: The Lagrange and Hermite interpolation functions for the plate bending elements ((ξ_i, ζ_i) denote the coordinates of the i–th node of the element, and (x_c, y_c) are the global coordinates of the center of the element).

Element type	Interpolation functions	Remarks
Lagrange elements:		
Linear	$\frac{1}{4}(1+\xi\xi_i)(1+\eta\eta_i)$	Node i ($i = 1, \ldots, 4$)
Quadratic	$\frac{1}{4}\xi\xi_i(1+\xi\xi_i)\eta\eta_i(1+\eta\eta_i)$	Corner node
	$\frac{1}{2}\eta\eta_i(1+\eta\eta_i)(1-\xi^2)$	Side node, $\xi_i = 0$
	$\frac{1}{2}\xi\xi_i(1+\xi\xi_i)(1-\eta^2)$	Side node, $\eta_i = 0$
	$(1-\xi^2)(1-\eta^2)$	Interior node
Seredipity element:		
Quadratic	$\frac{1}{4}(1+\xi\xi_i)(1+\eta\eta_i)(\xi\xi_i+\eta\eta_i-1)$	Corner node
	$\frac{1}{2}(1-\xi^2)(1+\eta\eta_i)$	Side node, $\xi_i = 0$
	$\frac{1}{2}(1+\xi\xi_i)(1-\eta^2)$	Side node, $\eta_i = 0$
Hermite cubic element:		
Interpolation functions for		
variable u	$\frac{1}{16}(\xi+\xi_i)^2(\xi\xi_i-2)(\eta+\eta_i)^2(\eta\eta_i-2)$	For node i ($i = 1, \ldots, 4$)
derivative $\partial u/\partial\xi$	$-\frac{1}{16}\xi_i(\xi+\xi_i)^2(\xi\xi_i-1)(\eta+\eta_i)^2(\eta\eta_i-2)$	
derivative $\partial u/\partial\eta$	$-\frac{1}{16}(\xi+\xi_i)^2(\xi\xi_i-2)\eta_i(\eta+\eta_i)^2(\eta\eta_i-1)$	
derivative $\partial^2 u/\partial\xi\,\partial\eta$	$\frac{1}{16}\xi_i(\xi+\xi_i)^2(\xi\xi_i-1)\eta_i(\eta+\eta_i)^2(\eta\eta_i-1)$	
Interpolation functions for		
variable u	$\frac{1}{2}(\xi_0+1)(\eta_0+1)(2+\xi_0+\eta_0-\xi^2-\eta^2)$	For node i ($i = 1, \ldots, 4$)
derivative $\partial u/\partial\xi$	$\frac{1}{2}\xi_i(\xi_0+1)^2(\xi_0-1)(\eta_0+1)$	
derivative $\partial u/\partial\eta$	$\frac{1}{2}\eta_i(\xi_0+1)(\eta_0+1)^2(\eta_0-1)]$	
	$\xi = (x-x_c)/a, \quad \eta = (y-y_c)/b$ $\xi_0 = \xi\xi_i, \quad \eta_0 = \eta\eta_i$	($2a$ and $2b$ are the sides of the rectangular element)

This choice of interpolation of (u_0, v_0, w_0) guarantees inter–element continuity of u_0, v_0, w_0, and $\frac{\partial w_0}{\partial n}$, and therefore such elements are called *conforming elements.* If the mixed derivative, $\frac{\partial^2 w_0}{\partial x \partial y}$, is not included in the interpolation, the resulting element violates the inter–element continuity of the derivative of the transverse displacement. Hence, it is termed a *nonconforming element.* The conforming element has a total of six degrees of freedom per node, whereas the nonconforming element has a total of five degrees of freedom per node (see Figure 5.4–1).

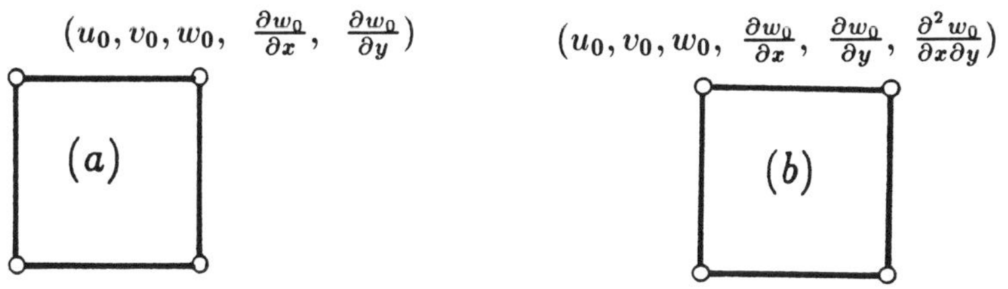

Figure 5.4–1 Plate bending elements of the classical plate theory.
(a) Nonconforming element $(u_0, v_0, w_0, \frac{\partial w_0}{\partial x}, \frac{\partial w_0}{\partial y})$.
(b) Conforming element $(u_0, v_0, w_0, \frac{\partial w_0}{\partial x}, \frac{\partial w_0}{\partial y}, \frac{\partial^2 w_0}{\partial x \partial y})$.

The finite element model of Eqs. (4.2–21)–(4.2–23) is (see Reddy [1] for details)

$$\sum_{\beta=1}^{3}\sum_{j=1}^{n_\beta}\left[\left(K_{ij}^{\alpha\beta}+G_{ij}^{\alpha\beta}\right)\Delta_j^\beta + M_{ij}^{\alpha\beta}\ddot{\Delta}_j^\beta\right] - F_i^\alpha = 0\ , \quad i = 1, 2, ..., n_\alpha \tag{5.4 – 3a}$$

or, in matrix form

$$([K^e] + [G^e])\{\Delta^e\} + [M^e]\{\ddot{\Delta}^e\} = \{F^e\} \tag{5.4 – 3b}$$

where $\alpha = 1, 2, 3$; $n_1 = n_2 = 4$ and $n_3 = 16$. The nodal values Δ_j^β, and coefficients $K_{ij}^{\alpha\beta}, G_{ij}^{\alpha\beta}$ and $M_{ij}^{\alpha\beta}$ are defined by

$$\Delta_j^1 = u_j\ ,\ \Delta_j^2 = v_j\ ,\ \Delta_j^3 = \Delta_j \tag{5.4 – 4}$$

and

$$K_{ij}^{1\alpha} = \int_{\Omega^e} \left(\frac{\partial \psi_i}{\partial x} N_{1j}^{\alpha} + \frac{\partial \psi_i}{\partial y} N_{6j}^{\alpha} \right) dxdy$$

$$K_{ij}^{2\alpha} = \int_{\Omega^e} \left(\frac{\partial \psi_i}{\partial x} N_{6j}^{\alpha} + \frac{\partial \psi_i}{\partial y} N_{2j}^{\alpha} \right) dxdy$$

$$K_{ij}^{3\alpha} = -\int_{\Omega^e} \left(\frac{\partial^2 \phi_i}{\partial x^2} M_{1j}^{\alpha} + 2\frac{\partial^2 \phi_i}{\partial x \partial y} M_{6j}^{\alpha} + \frac{\partial^2 \phi_i}{\partial y^2} M_{2j}^{\alpha} \right) dxdy$$

$$G_{ij}^{33} = \int_{\Omega^e} \left[\hat{N}_{xx} \frac{\partial \psi_i}{\partial x}\frac{\partial \psi_j}{\partial x} + \hat{N}_{xy} \left(\frac{\partial \psi_i}{\partial x}\frac{\partial \psi_j}{\partial y} + \frac{\partial \psi_i}{\partial x}\frac{\partial \psi_j}{\partial y} \right) + \hat{N}_{yy} \frac{\partial \psi_i}{\partial y}\frac{\partial \psi_j}{\partial y} \right] dxdy \qquad (5.4-5a)$$

$$M_{ij}^{11} = \int_{\Omega^e} I_0 \psi_i \psi_j dxdy \ , \ M_{ij}^{14} = \int_{\Omega^e} I_1 \psi_i \psi_j dxdy$$

$$M_{ij}^{22} = M_{ij}^{11} \ , \ M_{ij}^{25} = M_{ij}^{14} \ , \ M_{ij}^{33} = M_{ij}^{11}$$

$$M_{ij}^{44} = M_{ij}^{11} \ , \ M_{ij}^{45} = M_{ij}^{14} \ , \ M_{ij}^{55} = M_{ij}^{11} \qquad (5.4-5b)$$

$$F_i^1 = \int_{\Gamma^e} (N_{nn} - N_{nn}^T)\psi_i ds \ , \ F_i^2 = \int_{\Gamma^e} (N_{ns} - N_{ns}^T)\psi_i ds$$

$$F_i^3 = \int_{\Gamma^e} \left[\phi_i (V_n - V_n^T) + \frac{\partial \phi_i}{\partial n}(M_{nn} - M_{nn}^T) \right] ds - \int_{\Omega^e} \phi_i \ q dxdy \qquad (5.4-5c)$$

where

$$N_{1j}^1 = A_{11}\frac{\partial \psi_j}{\partial x} + A_{16}\frac{\partial \psi_j}{\partial y} \ , \ N_{1j}^2 = A_{16}\frac{\partial \psi_j}{\partial x} + A_{12}\frac{\partial \psi_j}{\partial y}$$

$$N_{1j}^3 = -\left(B_{11}\frac{\partial^2 \phi_j}{\partial x^2} + 2B_{16}\frac{\partial^2 \phi_j}{\partial x \partial y} + B_{12}\frac{\partial^2 \phi_j}{\partial y^2} \right)$$

$$N_{6j}^1 = A_{16}\frac{\partial \psi_j}{\partial x} + A_{66}\frac{\partial \psi_j}{\partial y} \ , \ N_{6j}^2 = A_{66}\frac{\partial \psi_j}{\partial x} + A_{26}\frac{\partial \psi_j}{\partial y}$$

$$N_{6j}^3 = -\left(B_{16}\frac{\partial^2 \phi_j}{\partial x^2} + 2B_{66}\frac{\partial^2 \phi_j}{\partial x \partial y} + B_{26}\frac{\partial^2 \phi_j}{\partial y^2} \right)$$

$$N_{2j}^1 = A_{12}\frac{\partial \psi_j}{\partial x} + A_{26}\frac{\partial \psi_j}{\partial y} \ , \ N_{2j}^2 = A_{26}\frac{\partial \psi_j}{\partial x} + A_{22}\frac{\partial \psi_j}{\partial y}$$

$$N_{2j}^3 = -\left(B_{12}\frac{\partial^2 \phi_j}{\partial x^2} + 2B_{26}\frac{\partial^2 \phi_j}{\partial x \partial y} + B_{22}\frac{\partial^2 \phi_j}{\partial y^2} \right)$$

$$
\begin{aligned}
M_{1j}^1 &= B_{11}\frac{\partial\psi_j}{\partial x} + B_{16}\frac{\partial\psi_j}{\partial y} \,, \quad M_{2j}^2 = B_{16}\frac{\partial\psi_j}{\partial x} + B_{12}\frac{\partial\psi_j}{\partial y} \\
M_{1j}^3 &= -\left(D_{11}\frac{\partial^2\phi_j}{\partial x^2} + 2D_{16}\frac{\partial^2\phi_j}{\partial x\partial y} + D_{12}\frac{\partial^2\phi_j}{\partial x^2}\right) \\
M_{6j}^1 &= B_{16}\frac{\partial\psi_j}{\partial x} + B_{66}\frac{\partial\psi_j}{\partial y} \,, \quad M_{6j}^2 = B_{66}\frac{\partial\psi_j}{\partial x} + B_{26}\frac{\partial\psi_j}{\partial y} \\
M_{6j}^3 &= -\left(D_{16}\frac{\partial^2\phi_j}{\partial x^2} + 2D_{66}\frac{\partial^2\phi_j}{\partial x\partial y} + D_{26}\frac{\partial^2\phi_j}{\partial y^2}\right) \\
M_{2j}^1 &= B_{12}\frac{\partial\psi_j}{\partial x} + B_{26}\frac{\partial\psi_j}{\partial y} \,, \quad M_{2j}^2 = B_{26}\frac{\partial\psi_j}{\partial x} + B_{22}\frac{\partial\psi_j}{\partial y} \\
M_{2j}^3 &= -\left(D_{12}\frac{\partial^2\phi_j}{\partial x^2} + 2D_{26}\frac{\partial^2\phi_j}{\partial x\partial y} + D_{22}\frac{\partial^2\phi_j}{\partial y^2}\right) \qquad (5.4-6)
\end{aligned}
$$

All other coefficients are zero. The stress and moment resultants, $V_n, N_{nn}, N_{ns}, M_{nn}$, and M_{ns}, on an edge with normal $\hat{\mathbf{n}} = n_x\hat{\mathbf{i}} + n_y\hat{\mathbf{j}}$ are defined in terms of the stress and moment resultants $(Q_x, Q_y, N_{xx}, N_{yy}, N_{xy}, M_{xx}, M_{yy}, M_{xy})$ as (see Figure 5.4–2)

$$
\begin{aligned}
N_{nn} &= n_x^2 N_{xx} + n_y^2 N_{yy} + 2n_x n_y N_{xy} \\
N_{ns} &= (N_{yy} - N_{xx})n_x n_y + N_{xy}(n_x^2 - n_y^2) \qquad (5.4-7a)
\end{aligned}
$$

$$
\begin{aligned}
M_{nn} &= n_x^2 M_{xx} + n_y^2 M_{yy} + 2n_x n_y M_{xy} \\
M_{ns} &= (M_{yy} - M_{xx})n_x n_y + M_{xy}(n_x^2 - n_y^2) \qquad (5.4-7b) \\
V_n &= n_x Q_x + n_y Q_y - \frac{\partial M_{ns}}{\partial s} \qquad (5.4-7c)
\end{aligned}
$$

Similar relations hold for the thermal resultants, V_n^T, N_{nn}^T, N_{ns}^T, M_{nn}^T, and M_{ns}^T.

The computation of the coefficients in Eqs. (4)–(6) requires evaluation of integrals over each element domain. When the element geometry is nonrectangular or the integrands are complicated functions of the coordinates (x, y), exact evaluation of the integrals by conventional methods is not possible. In finite element computer programs, such integrals are evaluated using numerical integrations methods, called *quadrature*.

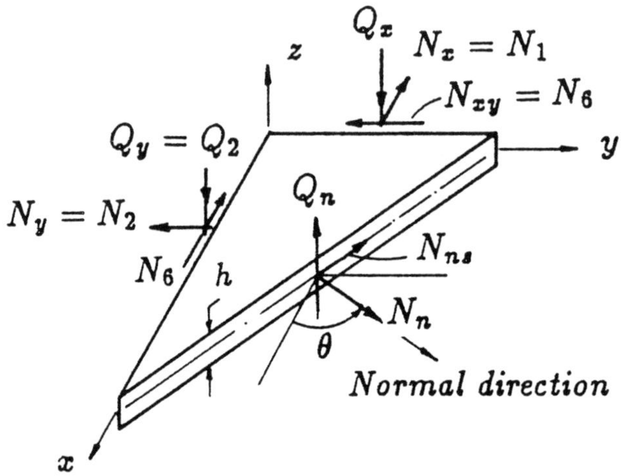

Figure 5.4–2: The stress resultants on an inclined edge.

Equations (3) of an element are assembled by using the interelement continuity of the nodal displacements and balance of nodal forces. Then the displacement and force boundary conditions of the problem are imposed on the assembled system of equations, and the equations are solved for nodal displacements of the total mesh. The strains in the problem coordinates (x, y, z) can be computed using the displacement field, and they can be transformed to lamina coordinates using the transformations in Eqs. (3.3–11). The stresses can be computed using the stress–strain relations (3.3–16).

5.4–2 First–Order Shear Deformation Theory

The weak form of the governing equations of the first–order theory contain, at the most, only the first derivatives of the dependent variables $(u_0, v_0, w_0, \phi_x, \phi_y)$. Therefore, they can all be approximated using the Lagrange interpolation functions

$$u_0 = \sum_{j=1}^{n} u_j(t)\psi_j(x,y)$$
$$v_0 = \sum_{j=1}^{n} v_j(t)\psi_j(x,y)$$
$$w_0 = \sum_{j=1}^{n} w_j(t)\psi_j(x,y) \qquad (5.4-8a)$$
$$\phi_x = \sum_{j=1}^{n} S_j^1\psi_j(x,y)$$
$$\phi_y = \sum_{j=1}^{n} S_j^2\psi_j(x,y) \qquad (5.4-8b)$$

where ψ_j are Lagrange interpolation functions. One can use linear, quadratic, or higher–order interpolation functions.

Substituting Eqs. (8) for $(u_0, v_0, w_0, \phi_x, \phi_y)$ into the weak form associated with Eqs. (4.3–14)–(4.3–18), we obtain the finite element model of the first–order theory

$$([K^e] + [G^e])\{\Delta^e\} + [M^e]\{\ddot{\Delta}^e\} = \{F^e\} \qquad (5.4-9)$$

where the coefficients are defined by $(\alpha = 1, 2, ..., 5)$

$$K_{ij}^{1\alpha} = \int_{\Omega^e} \left(\frac{\partial\psi_i}{\partial x}N_{1j}^{\alpha} + \frac{\partial\psi_i}{\partial y}N_{6j}^{\alpha}\right) \text{dxdy}$$
$$K_{ij}^{2\alpha} = \int_{\Omega^e} \left(\frac{\partial\psi_i}{\partial x}N_{6j}^{\alpha} + \frac{\partial\psi_i}{\partial y}N_{2j}^{\alpha}\right) dxdy$$
$$K_{ij}^{3\alpha} = \int_{\Omega^e} \left(\frac{\partial\phi_i}{\partial x}Q_{1j}^{\alpha} + \frac{\partial\phi_i}{\partial y}Q_{2j}^{\alpha}\right) dxdy$$
$$K_{ij}^{4\alpha} = \int_{\Omega^e} \left(\frac{\partial\psi_i}{\partial x}M_{1j}^{\alpha} + \frac{\partial\psi_i}{\partial y}M_{6j}^{\alpha} + \psi_i Q_{1j}^{\alpha}\right) dxdy$$
$$K_{ij}^{5\alpha} = \int_{\Omega^e} \left(\frac{\partial\psi_i}{\partial x}M_{6j}^{\alpha} + \frac{\partial\psi_i}{\partial y}M_{2j}^{\alpha} + \psi_i Q_{2j}^{\alpha}\right) dxdy \qquad (5.4-10)$$

The coefficients N^{α}_{Ij}, M^{α}_{Ij}, and Q^{α}_{Ij} for $\alpha = 1, 2, ..., 5$ and $I = 1, 2, 6$ are given by

$$N^1_{1j} = A_{11}\frac{\partial\psi_j}{\partial x} + A_{16}\frac{\partial\psi_j}{\partial y} \ , \ N^2_{1j} = A_{12}\frac{\partial\psi_j}{\partial y} + A_{16}\frac{\partial\psi_j}{\partial x}$$

$$N^4_{1j} = B_{11}\frac{\partial\psi_j}{\partial x} + B_{16}\frac{\partial\psi_j}{\partial y} \ , \ N^5_{1j} = B_{12}\frac{\partial\psi_j}{\partial y} + B_{16}\frac{\partial\psi_j}{\partial x}$$

$$N^1_{2j} = A_{12}\frac{\partial\psi_j}{\partial x} + A_{26}\frac{\partial\psi_j}{\partial y} \ , \ N^2_{2j} = A_{22}\frac{\partial\psi_j}{\partial y} + A_{26}\frac{\partial\psi_j}{\partial x}$$

$$N^4_{2j} = B_{12}\frac{\partial\psi_j}{\partial x} + B_{26}\frac{\partial\psi_j}{\partial y} \ , \ N^5_{2j} = B_{22}\frac{\partial\psi_j}{\partial y} + B_{26}\frac{\partial\psi_j}{\partial x}$$

$$N^1_{6j} = A_{16}\frac{\partial\psi_j}{\partial x} + A_{66}\frac{\partial\psi_j}{\partial y} \ , \ N^2_{6j} = A_{26}\frac{\partial\psi_j}{\partial y} + A_{66}\frac{\partial\psi_j}{\partial x}$$

$$N^3_{1j} = 0 \ , \ N^3_{2j} = 0 \ , \ N^3_{6j} = 0,$$

$$N^4_{6j} = B_{16}\frac{\partial\psi_j}{\partial x} + B_{66}\frac{\partial\psi_j}{\partial y} \ , \ N^5_{6j} = B_{26}\frac{\partial\psi_j}{\partial y} + B_{66}\frac{\partial\psi_j}{\partial x}$$

$$M^1_{1j} = B_{11}\frac{\partial\psi_j}{\partial x} + B_{16}\frac{\partial\psi_j}{\partial y} \ , \ M^2_{1j} = B_{12}\frac{\partial\psi_j}{\partial y} + B_{16}\frac{\partial\psi_j}{\partial x}$$

$$M^4_{1j} = D_{11}\frac{\partial\psi_j}{\partial x} + D_{16}\frac{\partial\psi_j}{\partial y} \ , \ M^5_{1j} = D_{12}\frac{\partial\psi_j}{\partial y} + D_{16}\frac{\partial\psi_j}{\partial x}$$

$$M^1_{2j} = B_{12}\frac{\partial\psi_j}{\partial x} + B_{26}\frac{\partial\psi_j}{\partial y} \ , \ M^2_{2j} = B_{22}\frac{\partial\psi_j}{\partial y} + B_{26}\frac{\partial\psi_j}{\partial x}$$

$$M^4_{2j} = D_{12}\frac{\partial\psi_j}{\partial x} + D_{26}\frac{\partial\psi_j}{\partial y} \ , \ M^5_{2j} = D_{22}\frac{\partial\psi_j}{\partial y} + D_{26}\frac{\partial\psi_j}{\partial x}$$

$$M^1_{6j} = B_{16}\frac{\partial\psi_j}{\partial x} + B_{66}\frac{\partial\psi_j}{\partial y} \ , \ M^2_{6j} = B_{26}\frac{\partial\psi_j}{\partial y} + B_{66}\frac{\partial\psi_j}{\partial x}$$

$$M^3_{1j} = 0 \ , \ M^3_{2j} = 0 \ , \ M^3_{6j} = 0$$

$$M^4_{6j} = D_{16}\frac{\partial\psi_j}{\partial x} + D_{66}\frac{\partial\psi_j}{\partial y} \ , \ M^5_{6j} = D_{26}\frac{\partial\psi_j}{\partial y} + D_{66}\frac{\partial\psi_j}{\partial x}$$

$$Q^3_{1j} = 0 \ , \ Q^3_{2j} = 0 \qquad (5.4-11)$$

$$
\begin{aligned}
F_i^1 &= \int_{\Gamma^e} \psi_i (N_{nn} - N_{nn}^T) ds \\
F_i^2 &= \int_{\Gamma^e} \psi_i (N_{ns} - N_{ns}^T) ds \\
F_i^3 &= \int_{\Omega^e} q\psi_i \; dxdy + \int_{\Gamma^e} \psi_i Q_n ds \\
F_i^4 &= \int_{\Gamma^e} \psi_i (M_{nn} - M_{nn}^T) ds \\
F_i^5 &= \int_{\Gamma^e} \psi_i (M_{ns} - M_{ns}^T) ds \qquad (5.4-12)
\end{aligned}
$$

When the four–node rectangular element is used, the element stiffness matrices are of the order 20 × 20; and for the nine-node element they are 45 × 45 (see Figure 5.4–3). If the inplane degrees of freedom (u_0, v_0) are omitted (for the uncoupled case), the stiffness matrices will be of orders 12 × 12 and 27 × 27 for the linear and quadratic rectangular elements, respectively [1,8–13]. The coefficients in Eq. (10)–(12) are evaluated, in general, using numerical integration methods.

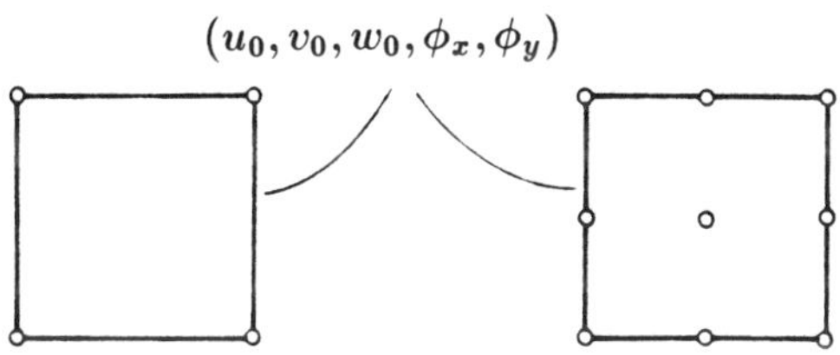

Figure 5.4–3: Plate bending elements of the first–order plate theory. Each node has the following degrees of freedom: $(u_0, v_0, w_0, \phi_x, \phi_y)$.

Once the finite element equations of all elements are available, they are put together by using the continuity of the nodal displacements and the balance of nodal forces. Then the boundary conditions on displacements and/or forces are applied on the assembled system of equations, and the equations are solved for the nodal values of the displacements. These nodal values can be used to determine the displacements, strains, and stresses at any point inside the element Ω^e, as discussed in more detail in Section 5.5.

5.4–3 Bending, Stability, and Vibration Models

The finite element models (5.4–3) and (5.4–9) of both classical and first–order shear deformation theories have the same form

$$[K^e]\{\Delta^e\} + [M^e]\{\ddot{\Delta}^e\} = \{F^e\} \qquad (5.4-13)$$

Equations (13) represent a system of ordinary differential equations in time in terms of the nodal values. These equations reduce to algebraic equations in the case of static, stability (or buckling), and natural vibration problems. For transient problems, the differential equations in (13) are reduced to algebraic equations, for each time step, by approximating the time derivatives.

Static Analysis

For the static bending analysis, we set the acceleration vector, $\{\ddot{\Delta}\}$, and the inplane applied forces to zero and obtain the equations

$$[K^e]\{\Delta^e\} = \{F^e\} \qquad (5.4-14)$$

Stability Analysis

For stability analysis (*i.e.*, to determine the value of inplane compressive forces at which the plate buckles; see Figure 5.4–4), we set the time derivative and force terms to zero and obtain

$$([K^e] - \lambda[S^e])\{\Delta^e\} = \{0\} \qquad (5.4-15)$$

where λ denotes the ratio of inplane forces and $[S^e]$ is the stability matrix, $[G^e] = -\lambda[S^e]$. For the first–order theory, for example, the stability (or geometric stiffness) matrix $[S^e]$ is given by

$$[S^e] = \begin{bmatrix} [0] & [0] & [0] & [0] & [0] \\ & [0] & [0] & [0] & [0] \\ & & [S^{33}] & [0] & [0] \\ & & & [0] & [0] \\ & & & [0] & [0] \\ & sym. & & & [0] \end{bmatrix} \tag{5.4-16}$$

where

$$S_{ij}^{33} = \int_{\Omega^e} \left[\bar{N}_{xx} \frac{\partial \psi_i}{\partial x} \frac{\partial \psi_j}{\partial x} + \bar{N}_{xy} \left(\frac{\partial \psi_i}{\partial x} \frac{\partial \psi_j}{\partial y} + \frac{\partial \psi_i}{\partial y} \frac{\partial \psi_j}{\partial x} \right) + \bar{N}_{yy} \frac{\partial \psi_i}{\partial y} \frac{\partial \psi_j}{\partial y} \right] dxdy \tag{5.4-17}$$

The eigenvalue λ in this case represents the ratio of actual buckling load to the applied inplane forces:

$$\lambda = \frac{N_{xx}}{\bar{N}_{xx}} = \frac{N_{yy}}{\bar{N}_{yy}} = \frac{N_{xy}}{\bar{N}_{xy}} \tag{5.4-18}$$

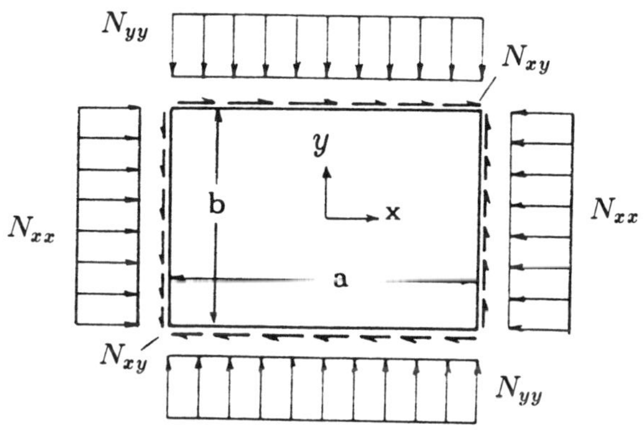

Figure 5.4 4: A rectangular plate element with applied inplane compressive and shear forces.

Natural Vibration Analysis

For natural vibration problems, we assume periodic motion and replace the acceleration vector by

$$\{\ddot{\Delta}^e\} = -\omega^2\{\Delta_0^e\} \quad (\text{or } \{\Delta^e\} = \{\Delta_0^e\}e^{-i\omega t}) \tag{5.4-19}$$

where ω denotes the frequency of natural vibration. Then Eq. (13) takes the form, under known applied in–plane forces, of an eigenvalue problem

$$(([K^e] + [G^e]) - \omega^2[M^e])\{\Delta_0^e\} = \{0\} \tag{5.4-20}$$

When the applied inplane forces are zero, $[G] = [0]$.

5.4–4 Transient Analysis

For the transient response, the time derivatives of the solution $\{\Delta\}$ in Eq. (13) should be approximated in terms of the solution itself to obtain algebraic equations relating the solution $\{\Delta\}$ at time $(t + \Delta t)$ to the solution at time t, where Δt denotes the time step. Such a procedure is called a time approximation scheme. The Newmark family of time approximation schemes is the most popular one in structural dynamics. In this scheme, the displacement and velocity vectors $(\{\Delta\}, \{\dot{\Delta}\})$ at time $t = (s+1)\Delta t$ are approximated by the expressions

$$\{\dot{\Delta}\}_{s+1} = \{\dot{\Delta}\}_s + [(1-\alpha)\{\ddot{\Delta}\}_s + \alpha\{\ddot{\Delta}\}_{s+1}]\Delta t \tag{5.4-21a}$$

$$\{\Delta\}_{s+1} = \{\Delta\}_s + \{\dot{\Delta}\}_s\Delta t + [(1-\gamma)\{\ddot{\Delta}\}_s + \gamma\{\ddot{\Delta}\}_{s+1}]\,\frac{(\Delta t)^2}{2} \tag{5.4-21b}$$

where s denotes the time step number, α and γ are parameters that control the accuracy and stability of the scheme, and the subscript s indicates that the vectors are evaluated at the s–th time step (*i.e.*, at time, $t = s\Delta t$).

Using Eqs. (13) and (21), we obtain the following algebraic equations for an element (see Reddy [1,6])

$$[\hat{K}]_{s+1}\{\Delta\}_{s+1} = \{\hat{F}\}_{s,s+1} \tag{5.4 – 22a}$$

where $\{\Delta\}_{s+1}$ denotes the value of $\{\Delta\}$ at time $t = (s+1)\delta t$, and

$$\begin{aligned}
[\hat{K}]_{s+1} &= [K]_{s+1} + a_1[M]_{s+1} \\
\{\hat{F}\}_{s,s+1} &= \{F\}_{s+1} + [M]_{s+1}(a_1\{\Delta\}_s + a_2\{\dot{\Delta}\}_s + a_3\{\ddot{\Delta}\}_s) \\
a_1 &= \frac{2}{\gamma(\Delta t)^2}\ ,\ a_2 = a_1\Delta t\ ,\ a_3 = \frac{1}{\gamma} - 1
\end{aligned} \tag{5.4 – 22b}$$

Once the solution $\{\Delta\}$ is known at time $t_{s+1} = (s+1)\Delta t$, the first and second derivatives of $\{\Delta\}$ (velocity and accelerations) at t_{s+1} can be computed from

$$\{\ddot{\Delta}\}_{s+1} = a_1(\{\Delta\}_{s+1} - \{\Delta\}_s) - a_2\{\dot{\Delta}\}_s - a_3\{\ddot{\Delta}\}_s \tag{5.4 – 23a}$$

$$\{\dot{\Delta}\}_{s+1} = \{\dot{\Delta}\}_s + a_4\{\ddot{\Delta}\}_s + a_5\{\ddot{\Delta}\}_{s+1} \tag{5.4 – 23b}$$

$$a_4 = (1-\alpha)\Delta t\ ,\quad a_5 = \alpha\Delta t \tag{5.4 – 23c}$$

Note that Eq. (22a) is valid for an element. Therefore, operations indicated in Eq. (22b) are carried out for an element, and $[\hat{K}^e]$ and $\{\hat{F}^e\}$ are assembled as in a static analysis. They are solved, after imposing the boundary conditions, for the solution at time, t_{s+1}. This procedure is repeated for any number of time steps. Thus, Eq. (22) represents a time–marching scheme.

For the first time step, the initial conditions on $\{\Delta(t)\}$ and $\{\dot{\Delta}(t)\}$ are used to compute $\{\Delta\}_0$ and $\{\dot{\Delta}\}_0$ for each element of the entire mesh. The acceleration vector $\{\ddot{\Delta}\}_0$ is computed from Eq. (13) at $t = 0$:

$$\{\ddot{\Delta}\}_0 = [M]^{-1}(\{F\}_0 - [K]\{\Delta\}_0) \tag{5.4 – 24}$$

Often it is assumed that $\{F\}_0 = \{0\}$. If the initial conditions are zero $\{\Delta\}_0 = \{0\}$ and the applied force is assumed to be zero at $t = 0$, we then take $\{\ddot{\Delta}\}_0 = \{0\}$.

Since Eq. (22) represents an approximation of the actual problem, error is introduced into the solution $\{\Delta\}_{s+1}$. This error can grow from one time step to the next as we solve Eq. (22a) for different times. If the error grows unboundedly with time, such time–approximation schemes are said to be *unstable.* If the error is bounded for all times, the scheme is said to be *stable.* For some schemes, the error is bounded only if certain restrictions on the time step are met. Such schemes are called *conditionally stable schemes.* Of course, we only use either stable or conditionally stable schemes.

The numerical stability of the Newmark scheme depends on the choice of the values for α and γ. Only when $\alpha = \gamma = 1/2$, is the scheme unconditionally stable. For other choices of these parameters, the time increment Δt must be chosen such that the following *stability criterion* is met:

$$\Delta t < \Delta t_{cr} = \left[\frac{1}{2}\omega_{\max}^2\left(\alpha - \gamma\right)\right]^{-\frac{1}{2}} , \; for\; \alpha \geq \frac{1}{2} , \; \gamma < \alpha \qquad (5.4-25)$$

where $\omega_{\max}$ denotes maximum natural frequency of the structure being analyzed, obtained with the same finite element mesh as that used for the transient analysis.

The parameters α and γ in the Newmark time approximation scheme (21) can be chosen to yield several well–known finite difference schemes. Some of them are listed here:

$$\begin{aligned}
&\alpha = \frac{1}{2} , \; \gamma = \frac{1}{2} : \text{ constant–average acceleration method} \\
&\alpha = \frac{1}{2} , \; \gamma = \frac{1}{3} : \text{ linear acceleration method} \\
&\alpha = \frac{1}{2} , \; \gamma = 0 : \text{ central difference method} \qquad (5.4-26)
\end{aligned}$$

The constant–average acceleration scheme is a stable scheme, and the linear acceleration scheme and central difference scheme are conditionally stable schemes.

5.4–5 Shear locking

The displacement–based C^0–plate elements (*i.e.*, those with no derivatives included as generalized displacements) of the first–order shear deformation plate theory are among the simplest available in the literature. Unfortunately, the accuracy of the low–order elements of this type (*i.e.*, those with nine or fewer nodes) is not satisfactory when the span–to–thickness ratio of the elements is large (*i.e.*, thin plates). For thin plates, the shearing strains ε_{xz} and ε_{yz} are required to vanish, and the plate elements based on the first–order theory become excessively stiff, yielding displacements that are too small compared to the true solution. This type of behavior, known as *shear locking*, has plagued shear deformable elements since their inception (see [16]).

Shear locking is due to the inability of shear deformable elements to model accurately the curvatures within an element under a state of zero transverse shearing strain. The various means employed to alleviate shear locking in plate elements include reduced integration [17–19], mixed formulations [20,21], hybrid formulations [22], assumed strain techniques [23], and discrete Kirchhoff methods [24], to name only a few. Of the large number of new displacement–based finite elements for the first–order shear deformation theory, in the literature, only a few have been reasonably successful [19–21], and there is still much room for improvement in terms of accuracy, ease of implementation, and computational efficiency.

A better understanding of locking phenomena is necessary in order to give guidelines for the development of new elements and to allow a more complete analysis of existing elements. When thin plates are modeled by plate finite elements with shear deformation, the energy due to transverse shear strains must vanish. Numerically this is equivalent to a constraint condition that requires that the product of the shear stiffness matrix and the displacement vector be zero. Therefore, in order to obtain a nontrivial solution, the shear stiffness matrix must be singular. One way to achieve the singularity of the transverse shear stiffness matrix is to use an order of numerical integration lower than is necessary to evaluate the integrals exactly. Thus, *reduced integration* of transverse shear stiffnesses is necessary. These observations led to explanations of locking behavior based upon measures of singularity of the shear stiffness matrix [25,26], the number of independent shear constraints imposed on an element in the thin plate limit, and other criteria. Higher–order elements or refined meshes of lower–order elements experience relatively less locking, but sometimes at the expense of rate of convergence.

5.5 COMPUTATION OF STRESSES AND STRAINS

Once the nodal values of generalized displacements have been obtained by solving the assembled equations of a problem, the strains are evaluated in each element by differentiating the displacements. Since, the displacements are only continuous across the element boundaries, strain continuity across the boundaries is not ensured. That is, along a boundary common to two elements, the strains and, hence, stresses take different values on the two sides of the interface. However, strains and, hence, stresses are continuous within an element.

Although the strains can be computed at any point in the element, including the nodes, Barlow [27,28] showed that the strains and stresses are the most accurate if they are computed at the $(N-1) \times (N-1)$ Gauss points, where $N \times N$ is the Gauss quadrature rule used to evaluate the stiffness coefficients. This reduced order integration rule is termed as the *reduced integration*, and the reduced integration points are called the Barlow points. For example, the linear rectangular plate bending element of the first-order theory requires 2×2 integration to evaluate the bending stiffnesses exactly. Then the one–point integration (or 1×1) should be used to evaluate the strains and stresses. Similarly, for a quadratic rectangular element, the Barlow points are at the Gauss points of the 2×2 Gauss rule.

Note that the displacements in the finite element models are referred to the global coordinates (x, y, z). Differentiation of these values at the Barlow points yields the strains there in the global coordinates. However, we require stresses and strains in the lamina (or principal material) coordinates to check for failures. In order to obtain strains and stresses in the material coordinates, we need to transform the strains and stresses to material coordinates associated with each layer. First the global strains ε_i are transformed to the lamina strains $\bar{\varepsilon}_i$ using the strain transformation equations [see Eq. (3.3–11)]:

$$\begin{Bmatrix} \varepsilon_1 \\ \varepsilon_2 \\ \varepsilon_3 \\ \varepsilon_4 \\ \varepsilon_5 \\ \varepsilon_6 \end{Bmatrix} = [R] \begin{Bmatrix} \varepsilon_{xx} \\ \varepsilon_{yy} \\ \varepsilon_{zz} \\ \gamma_{yz} \\ \gamma_{xz} \\ \gamma_{xy} \end{Bmatrix} \tag{5.5 – 1a}$$

where $[R]$ denotes the matrix

$$\begin{bmatrix} \cos^2\theta & \sin^2\theta & 0 & 0 & 0 & \sin\theta\cos\theta \\ \sin^2\theta & \cos^2\theta & 0 & 0 & 0 & -\sin\theta\cos\theta \\ 0 & 0 & 1 & 0 & 0 & 0 \\ 0 & 0 & 0 & \cos\theta & -\sin\theta & 0 \\ 0 & 0 & 0 & \sin\theta & \cos\theta & 0 \\ -2\sin\theta\cos\theta & 2\sin\theta\cos\theta & 0 & 0 & 0 & \cos^2\theta - \sin^2\theta \end{bmatrix} \tag{5.5 – 1b}$$

Then the lamina constitutive equations are used to compute the stresses, referred to the material coordinates, at the Barlow points:

$$\begin{Bmatrix} \sigma_1 \\ \sigma_2 \\ \sigma_6 \end{Bmatrix} = \begin{bmatrix} Q_{11} & Q_{12} & 0 \\ Q_{12} & Q_{22} & 0 \\ 0 & 0 & Q_{66} \end{bmatrix} \begin{Bmatrix} \varepsilon_1 \\ \varepsilon_2 \\ \varepsilon_6 \end{Bmatrix} \tag{5.5 – 2a}$$

$$\begin{Bmatrix} \sigma_4 \\ \sigma_5 \end{Bmatrix} = \begin{bmatrix} Q_{44} & 0 \\ 0 & Q_{55} \end{bmatrix} \begin{Bmatrix} \varepsilon_4 \\ \varepsilon_5 \end{Bmatrix} \tag{5.5 – 2b}$$

Note that $[Q]$ matrix represents the matrix of material stiffnesses associated with the principal coordinates of the layer (lamina) as discussed in Section 3.4.

While the inplane stresses $(\sigma_{xx}, \sigma_{yy}, \sigma_{xy})$ can be computed only through stress–strain (*i.e.*, constitutive) relations, the transverse stresses, known as the *interlaminar stresses* $(\sigma_{xz}, \sigma_{yz}, \sigma_{zz})$ can be computed through either the constitutive equations (2b) or the equilibrium equations of 3–D elasticity

$$\frac{\partial \sigma_{xx}}{\partial x} + \frac{\partial \sigma_{xy}}{\partial y} + \frac{\partial \sigma_{xz}}{\partial z} = 0 \qquad (5.5-3a)$$

$$\frac{\partial \sigma_{xy}}{\partial x} + \frac{\partial \sigma_{yy}}{\partial y} + \frac{\partial \sigma_{yz}}{\partial z} = 0 \qquad (5.5-3b)$$

$$\frac{\partial \sigma_{xz}}{\partial x} + \frac{\partial \sigma_{yz}}{\partial y} + \frac{\partial \sigma_{zz}}{\partial z} = 0 \qquad (5.5-3c)$$

Integration of Eqs. (3a,b) with respect to z give $(\sigma_{xz}, \sigma_{yz})$, which can be substituted into Eq. (3c) to determine σ_{zz} (see Reddy [4]). It is found that for homogeneous plates and thin laminates the stresses computed from the equilibrium equations (3) are found to be very accurate.

5.6 EXAMPLE PROBLEMS

5.6–1 General Comments

Here we present numerical results for a number of example problems of bending, vibration, and buckling of rectangular laminated plates. For the purpose of comparisons, the following two sets of lamina properties, typical of graphite–epoxy material, are used in the numerical examples unless stated otherwise:

$$E_1 = 25E_2, \quad G_{12} = G_{13} = 0.5E_2, \quad G_{23} = 0.2E_2, \quad \nu_{12} = 0.25 \qquad (5.6-1)$$

$$E_1 = 40E_2, \quad G_{12} = G_{13} = 0.6E_2, \quad G_{23} = 0.5E_2, \quad \nu_{12} = 0.25 \qquad (5.6-2)$$

The shear correction coefficient for the first–order theory is taken to be $K = 5/6$. In the finite–element analysis a 4×4 mesh ($m \times n$ mesh means m elements along the x–coordinate and n elements along the y–coordinate) is used for the classical theory, while 2×2 mesh of nine–node quadratic elements is used for the first–order theory when a quadrant of a plate is analyzed (for SS, CC and FF boundary conditions), and equivalent 8×4 and 4×2 meshs are used for the half–plate models (in the case of SC, FS and FC boundary conditions). The notation SC, for example, refers to the boundary conditions used on the edges $x = \pm a/2$, while the other two edges (*i.e.*, $y = 0, b$), in all cases, are simply supported. The numerical results presented here are taken from the papers of Reddy and his colleagues [1,6,8,31]. For additional numerical results, the readers may consult the papers of Khdeir and his colleagues [34–46].

5.6–2 Bending Analysis

The loading in all cases is assumed to be sinusoidal (see Figure 5.2–1 for the geometry of the plate)

$$q(x, y) = q_0 \cos \frac{\pi x}{a} \sin \frac{\pi y}{b} \qquad (5.6-3)$$

In the tables and figures, the results for deflections and stresses are presented using the following nondimensional form:

$$\overline{w} = \frac{w_0(0, b/2) E_2 h^3}{b^4 q_0} 10^2 \qquad (5.6-4)$$

$$\overline{\sigma}_{xx} = \frac{\sigma_{xx}(0, \frac{b}{2}, -\frac{h}{2}) h^2}{b^2 q_0} \times 10$$

$$\overline{\sigma}_{yy} = \frac{\sigma_{yy}(0, \frac{b}{2}, \frac{h}{2}) h^2}{b^2 q_0} \times 10$$

$$\overline{\sigma}_{yz} = \frac{\sigma_{yz}(0, 0, 0) h}{b q_0} \times 10 \qquad (5.6-5)$$

where h is the total thickness of the laminate.

Figures 5.6–1 and 5.6–2 contain plots of deflections versus side–to–thickness ratio b/h of two–layer and ten–layer antisymmetric $(0/90/\ldots)_{even}$ cross–ply laminates with various boundary conditions. The classical laminate theory always underpredicts deflections because the plate is modeled as infinitely stiff through the thickness. Figures 5.6–3 and 5.6–4 contain plots of $\overline{w}$ vs. E_1/E_2. As the degree of orthotropy increases, the difference between the deflections predicted by the classical and the first–order shear deformation theories increases, indicating that the shear deformation effect is more significant in anisotropic plates.

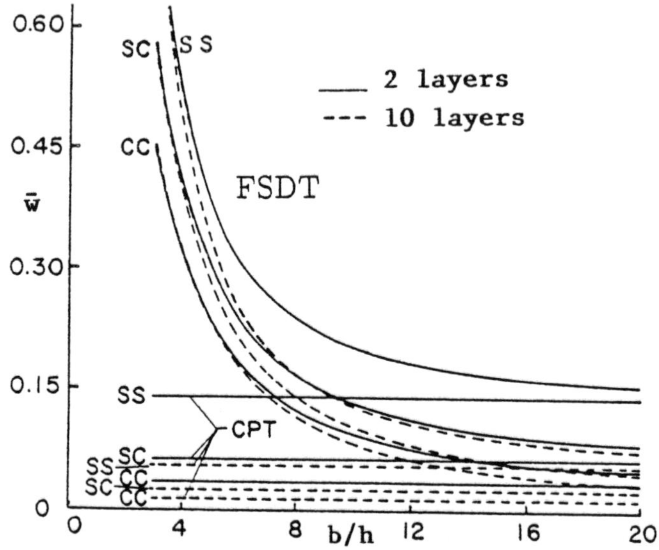

Figure 5.6–1: Nondimensionalized center deflection versus side–to–thickness ratio for SS, SC, and CC boundary conditions ($a = b/2$).

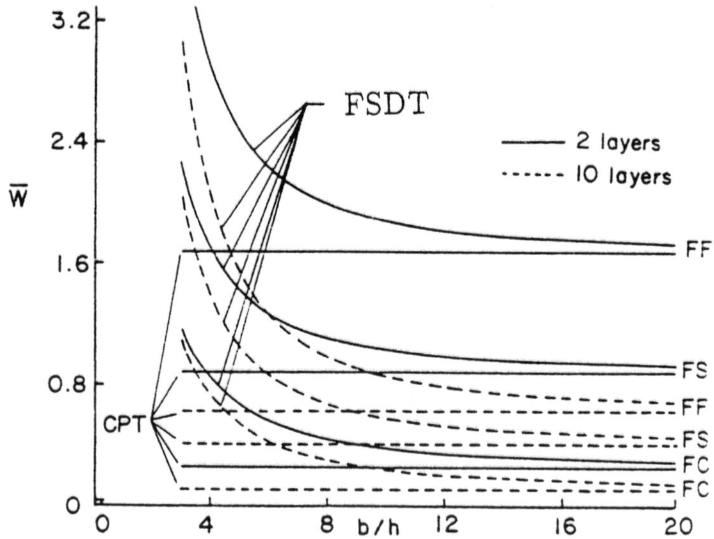

Figure 5.6–2: Nondimensionalized center deflection versus side–to–thickness ratio for FF, FS, and FC boundary conditions ($a = b/2$).

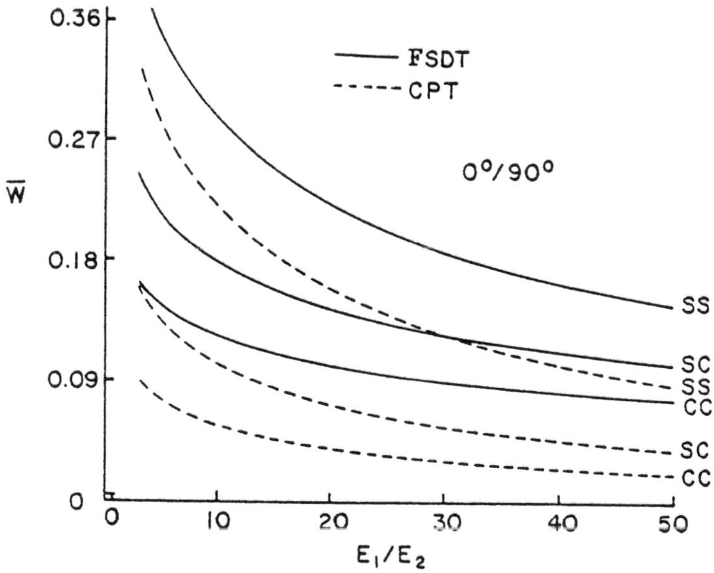

Figure 5.6–3: Nondimensionalized center deflection versus modulus ratio E_1/E_2 for SS, SC, and CC boundary conditions $(b/h = 10, a = b/2)$.

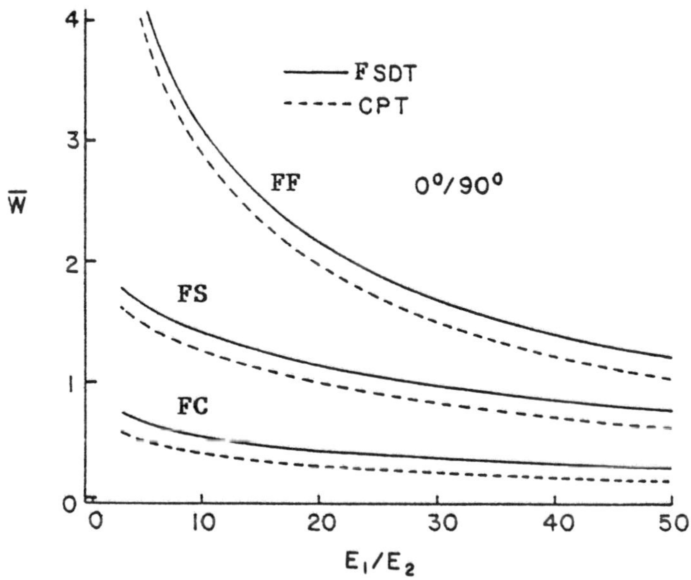

Figure 5.6–4: Nondimensionalized center deflection versus modulus ratio E_1/E_2 for FF, FS, and FC boundary conditions $(b/h = 10, a = b/2)$.

Tables 5.6–1 through 5.6–4 contain numerical values of deflections and stresses obtained using analytical (first row) and finite element models (second row). Quarter plate models with 2×2 mesh are used for SS, CC and FF boundary conditions, and half plate models with 4×2 mesh are used for all other boundary conditions. In all cases the finite element results are in reasonable agreement with the exact solutions. Of course, use of refined meshes would improve the accuracy of the finite element solutions.

Table 5.6–1: Nondimensional center deflection ($\overline{w}$) of antisymmetric cross–ply square plates, including various boundary conditions.

No. of Layers	$\frac{b}{h}$	Theory	SS	SC	CC	FF	FS	FC
	5	FSDT	1.758	1.477	1.257	2.777	2.335	1.897
			1.759	1.478	1.257	2.776	2.334	1.897
		CLPT	1.064	0.664	0.429	1.777	1.471	0.980
2			1.043	0.648	0.417	1.786	1.465	0.977
	10	FSDT	1.237	0.883	0.656	2.028	1.687	1.223
			1.238	0.883	0.657	2.027	1.687	1.223
		CLPT	1.064	0.664	0.429	1.777	1.471	0.980
			1.043	0.648	0.417	1.786	1.465	0.977
	5	FSDT	1.137	1.045	0.945	1.663	1.460	1.258
			1.137	1.045	0.945	1.662	1.460	1.258
		CLPT	0.442	0.266	0.167	0.665	0.579	0.380
10			0.444	0.266	0.169	0.686	0.593	0.391
		FSDT	0.615	0.480	0.385	0.915	0.800	0.612
	10		0.616	0.480	0.386	0.914	0.800	0.612
		CLPT	0.442	0.266	0.167	0.665	0.579	0.380
			0.444	0.266	0.169	0.686	0.593	0.391

Table 5.6–2: Nondimensional axial stress ($\overline{\sigma}_{xx}$) of antisymmetric cross–ply square plates, including various boundary conditions.

No. of Layers	$\frac{b}{h}$	Theory	SS	SC	CC	FF	FS	FC
	5	FSDT	7.157	5.338	3.911	2.469	4.430	2.434
			6.948	5.465	3.707	2.359	4.479	2.542
		CLPT	7.157	5.660	4.800	2.403	4.442	3.042
2			6.659	5.782	4.348	2.034	4.288	2.991
	10	FSDT	7.157	5.494	4.450	2.442	4.435	2.790
			6.948	5.668	4.222	2.331	4.491	2.895
		CLPT	7.157	5.660	4.800	2.403	4.442	3.042
			6.659	5.782	4.348	2.034	4.288	2.991
	5	FSDT	5.009	3.707	2.275	1.712	2.957	1.343
			4.864	3.755	2.154	1.639	2.993	1.427
		CLPT	5.009	3.829	3.167	1.725	2.986	1.865
10			4.611	3.911	2.798	1.324	2.806	1.700
	10	FSDT	5.009	3.642	2.692	1.723	2.968	1.594
			4.863	3.754	2.550	1.648	3.010	1.674
		CLPT	5.009	3.829	3.167	1.725	2.986	1.865
			4.611	3.911	2.798	1.324	2.806	1.700

Deflection and stresses obtained with the first–order theory are compared with the 3–D elasticity solutions in Table 5.6–5 for five–layer ($h_1 = h_3 = h_5 = h/6, h_2 = h_4 = h/4$) cross–ply ($0°/90°/0°/90°/0°$) square plate under sinusoidal loading. The finite element solution was obtained using the 2×2 mesh of eight–node quadratic elements in a quarter plate. The material properties used are those in Eq. (1).

Table 5.6–3: Nondimensional axial stress ($\overline{\sigma}_{yy}$) of antisymmetric cross–ply square plates, including various boundary conditions.

Layers	$\frac{b}{h}$	Theory	SS	SC	CC	FF	FS	FC
	5	FSDT	7.157	6.034	5.153	11.907	9.848	8.047
			6.948	5.1914	4.990	11.675	9.140	8.367
		CLPT	7.157	4.483	2.914	11.849	9.837	6.560
2			6.659	4.393	2.615	11.614	8.878	7.181
	10	FSDT	7.157	5.109	3.799	11.884	9.847	7.150
			6.948	5.082	3.661	11.654	9.1201	7.610
		CLPT	7.157	4.483	2.914	11.849	9.837	6.560
			6.659	4.393	2.615	11.614	8.878	7.181
	5	FSDT	5.009	4.628	4.212	7.583	6.590	5.706
			4.864	4.511	4.086	7.429	6.141	5.844
		CLPT	5.009	3.025	1.911	7.480	6.531	4.284
10			4.611	2.942	1.694	7.395	5.935	4.782
	10	FSDT	5.009	3.904	3.135	7.533	6.566	5.029
			4.863	3.852	3.031	7.384	6.097	5.279
		CLPT	5.009	3.025	1.911	7.480	6.531	4.284
			4.611	2.942	1.694	7.395	5.935	4.782

Table 5.6–4: Nondimensional transverse shear stress ($\overline{\sigma}_{yz}$) of antisymmetric cross–ply square plates, including various boundary conditions (FSDT, exact solution).

Layers	b/h	SS	SC	CC	FF	FS	FC
	5	2.729	2.297	1.958	3.901	3.390	2.748
2							
	10	2.729	1.993	1.523	3.882	3.383	2.449
	5	2.729	2.498	2.248	3.883	3.437	2.951
10							
	10	2.729	2.126	1.708	3.853	3.421	2.605

Table 5.6–5: Comparison of maximum deflection and stesses for five–layer, cross–ply, square plate under sinusoidal loading.

a/h	Source	$\overline{w}$	$\pm\overline{\sigma}_{xx}$	$\pm\overline{\sigma}_{yy}$	$-\overline{\sigma}_{xy}$	$\overline{\sigma}_{xz}$	$\overline{\sigma}_{yz}$
	3-DES[a]	18.668	0.685	0.633	0.0394	0.238	0.229
4	Exact	15.623	0.425	0.489	0.0227	0.296	0.0932
	FEM	15.602	0.4241	0.4879	0.0228	0.296	0.0933
	3-DES	6.830	0.545	0.430	0.0246	0.258	0.223
10	Exact	6.213	0.488	0.400	0.0214	0.335	0.0755
	FEM	6.201	0.487	0.397	0.0214	0.336	0.0776
	3-DES	4.981	0.539	0.380	0.0222	0.268	0.212
20	Exact	4.796	0.513	0.365	0.0208	0.351	0.0714
	FEM	4.786	0.512	0.364	0.0209	0.351	0.0715
	3-DES	4.377	0.539	0.360	0.0213	0.272	0.205
100	Exact	4.332	0.524	0.350	0.0206	0.357	0.0688
	FEM	4.314	0.521	0.347	0.0206	0.361	0.0705
CLPT		4.3505	0.539	0.359	0.0213	0.272	0.205

[a] 3-DES=3–D elasticity solution of Pagano [30].

Table 5.6–6 contains the maximum deflection and stresses in a three-layer ($h_1 = h_3 = 0.1h$, $h_2 = 0.8h$), sandwich square plate ($0°/90°/0°$) subjected to sinusoidal loading. The material properties of the face sheets (*i.e.*, layers 1 and 3) are the same as in Eq. (1), and the core material is assumed to be transversely isotropic with respect to the $z-$coordinate and is characterized by the following properties (see Pagano [30]):

$$E_1 - E_2, \; E_3 - 12.5E_2, \; G_{13} - G_{23} - 1.5E_2, \; G_{12} = 0.4E_2 \tag{5.6 - 6}$$

It is clear that the effect of shear deformation is significant for

a/h smaller than 10. The finite element and exact solutions of the FSDT are in good agreement. It is also clear that FSDT does not yield accurate results for very thick plates.

Table 5.6–6: Comparison of maximum deflection and stresses in a three–layer sandwich (0°/90°/0°) plate subjected to sinusoidal loading ($h_1 = h_3 = 0.1h, h_2 = 0.8h$).

a/h	Source	$\bar{w}$	$\bar{\sigma}_{xx}$	$\bar{\sigma}_{yy}$	$\bar{\sigma}_{xy}$	$\bar{\sigma}_{xz}$	$\bar{\sigma}_{yz}$
	3-DES[a]	-	3.278	0.4517	0.2403	0.185	0.140
2	Exact	9.552	0.718	0.3622	0.1954	0.876	0.252
	FEM	9.543	0.716	0.3615	0.1961	0.877	0.253
	3-DES	-	1.556	0.2595	0.1437	0.239	0.107
4	Exact	4.7672	0.867	0.1520	0.0877	0.993	0.174
	FEM	4.7606	0.865	0.1517	0.0878	0.994	0.174
	3-DES	-	1.153	0.1104	0.0707	0.300	0.053
10	Exact	1.560	1.017	0.0776	0.0533	1.110	0.095
	FEM	1.5578	1.015	0.0774	0.0535	0.112	0.095
	3-DES	-	1.110	0.0700	0.0511	0.317	0.036
20	Exact	1.0524	1.053	0.0595	0.0450	1.140	0.076
	FEM	1.0503	1.051	0.0594	0.0452	1.140	0.076
	3-DES	-	1.064	0.0541	0.0425	0.323	0.031
50	Exact	0.9064	1.065	0.0539	0.0424	1.148	0.070
	FEM	0.9040	1.062	0.0538	0.0425	1.151	0.071
	3-DES	-	1.098	0.0550	0.0437	0.324	0.0297
100	Exact	0.8852	1.067	0.0531	0.0420	1.149	0.069
	FEM	0.882	1.063	0.0530	0.0421	1.158	0.072
CLPT		-	1.097	0.0534	0.0433	0.324	0.0295

[a] 3-DES=3–D elasticity solution of Pagano [30].

The effect of loading, stacking sequence, and shear deformation on the deflections of cross–ply laminates subjected to sinusoidal distribution of temperature alone or combined temperature and mechanical loading are presented in Table 5.6–7 (see [31]). The mechanical material properties used are again the same as those in Eq. (1). The temperature distribution is assumed to be of the form

$$T(x,y,z) = T_0(x,y) + \frac{z}{h}T_1(x,y) \qquad (5.5-7)$$

The thermal data used are: $T_0 = 0, T_1 = 1.0, \alpha_1 = 10^{-6}$, and $\alpha_2 = 3\alpha_1$, where α_i denote the coefficients of thermal expansion in the principal material coordinates. The first line for each value of a/h in the table corresponds to the exact solution and the second line to the finite element solution [$\overline{w} = 10hw/\alpha_1\overline{T}_1a^2$, $\hat{w} = w[(q_0a^4/h^3\beta) + (\alpha_1\overline{T}_1a^2/10h)]$, $\beta = \pi^4 4G_{12} + [E_1 + (1 + \nu_{12})E]/(1-\nu_{12}\nu_{21})/12]$, and $\alpha_2 = 3\alpha_1$.

Table 5.6–7: Effects of loading, lamination, and thickness on the nondimensionalized deflections for a simply supported (SS–1) cross–ply square plates.

	Sinusoidal temperature				Combined loading	
$\frac{a}{h}$	0°	0°/90°	0°/90°/0°		0°/90°	0°/90°/0°
100	1.0313	1.6765	1.0949	0.9962	2.4451	0.9973
	(1.0312)†	(1.6764)	(1.0948)	(0.9923)	(2.4284)	(0.9933)
50	1.0317	1.6765	1.0963	1.0105	2.4597	1.0138
	(1.0317)	(1.6764)	(1.0962)	(1.0079)	(2.4515)	(1.0112)
25	1.0334	1.6765	1.1018	1.0683	2.5083	1.0811
	(1.0333)	(1.6764)	(1.1017)	(1.0659)	(2.5022)	(1.0787)
20	1.0346	1.6765	1.1058	1.1115	2.5443	1.1312
	(1.0345)	(1.6764)	(1.1057)	(1.1091)	(2.5384)	(1.1288)
10	1.0440	1.6765	1.1365	1.4671	2.8438	1.5384
	(1.0439)	(1.6764)	(1.1364)	(1.4645)	(2.8380)	(1.5456)
5	1.0721	1.6765	1.2224	2.8332	4.0415	3.0377
	(1.0720)	(1.6764)	(1.2224)	(2.8292)	(4.0346)	(3.0333)

† - Finite element solution.

Table 5.6–8 shows the maximum nondimensionalized deflections of simply–supported (see Figure 5.6–5), two– and sixteen–layer angle–ply ($\theta/-\theta/\theta/-\theta,\ldots$) square plates under sinusoidal loading. The material properties used are those listed in Eq. (2).

Table 5.6–8: Nondimensionalized deflection as a function of number of layers, angle, and side–to–thickness ratio for simply–supported (SS–2) angle–ply square plates.

a/h	$\theta = 5^0$		$\theta = 30^0$		$\theta = 45^0$	
	$n = 2$	$n = 16$	$n = 2$	$n = 16$	$n = 2$	$n = 16$
2	37.850	36.840	33.325	29.878	32.300	29.131
4	13.164	12.596	12.155	8.875	11.575	8.429
5	9.760	9.256	9.568	6.316	9.088	5.938
6.25	7.475	7.008	7.909	4.669	7.496	4.346
10	4.883	4.454	6.099	2.872	5.773	2.621
12.5	4.264	3.843	5.678	2.456	5.376	2.224
20	3.585	3.172	5.224	2.005	4.944	1.793
25	3.427	3.105	5.119	1.900	4.844	1.693
50	3.215	2.806	4.979	1.761	4.711	1.560
100	3.162	2.754	4.944	1.726	4.678	1.527

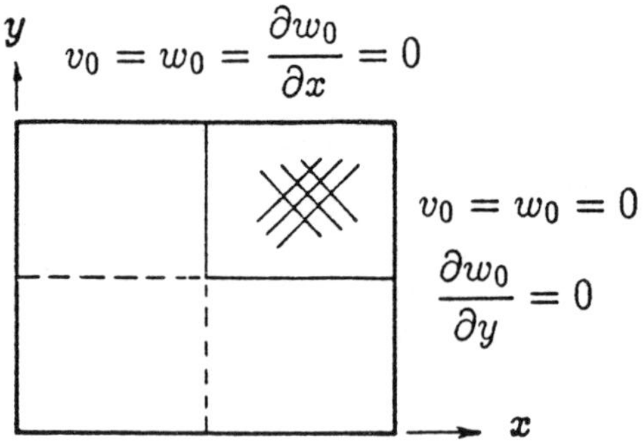

At a line of symmetry with normal **n**: $u_s = 0$

At an edge with normal **n**: $u_n = 0$ and w_0=0

Figure 5.6–5: Simply supported boundary conditions (SS–2) used for antisymmetric angle–ply laminates.

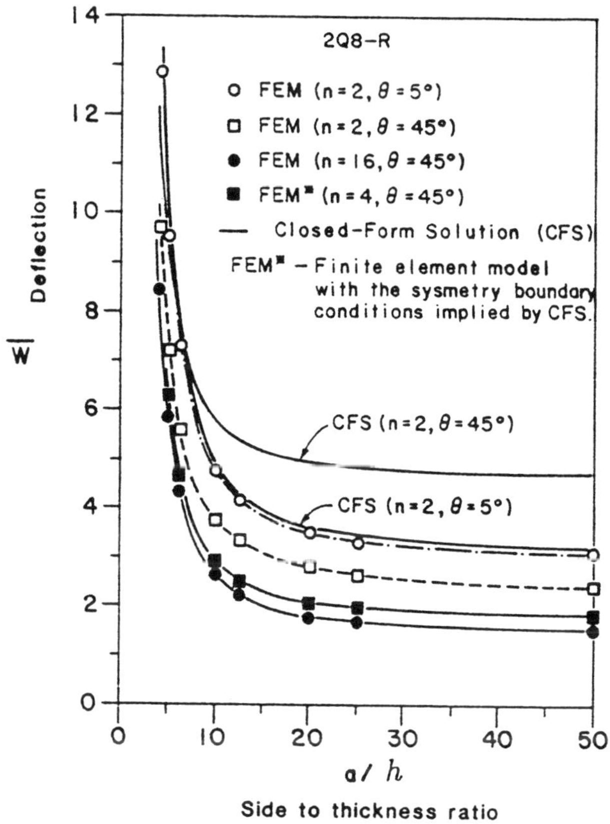

Figure 5.6–6: Nondimensionalized maximum deflection versus side–to–thickness ratio of angle–ply plates.

Figure 5.6–6 shows the effect of side–to–thickness ratio (a/h), number of layers, and stacking sequence on the nondimensionalized maximum deflection of angle–ply plates (from Reddy and Chao [8]). The material properties used are those listed in Eq. (2). The finite element results are based on the first shear deformation theory (FSDT), and were obtained using 2×2 mesh of eight–node quadratic elements in a quarter plate. The boundary conditions, $u_0(a/2, y) = v_0(x, b/2) = 0$, along the lines of symmetry of a quarter plate are valid for cross–ply laminates (SS–1). However, the same boundary conditions for antisymmetric angle–ply laminates (SS–2) are incorrect. The correct boundary conditions along the lines of biaxial symmetry are $v_0(a/2, y) = u_0(x, b/2) = 0$. The effect of using improper symmetry boundary conditions on the deflections is illustrated in the figure.

Table 5.6–9 contains nondimensionalized deflections $\overline{w} \equiv w(0, b/2)E_2h^3 \times 10^2/(a^4q_0)$ of angle–ply laminates subjected to uniformly distributed transverse load under various boundary conditions and with different values of E_1/E_2 ($G_{12} = G_{13} = 0.6E_2, G_{23} = 0.5E_2, \nu_{12} = 0.25$). The results were obtained using the Lévy method with state–space approach discussed in Section 5.3 (see Khdeir [43]).

Table 5.6–9: Nondimensionalized deflections of simply supported (SS–2), four–layer antisymmetric angle–ply square plates ($45°/-45°/45°/-45°, a/h = 10$).

Theory	$\frac{E_1}{E_2}$	SS	SC	CC	FF	FS	FC
FSDT	2	3.3749	2.4227	1.7530	10.7347	6.4469	4.7427
CLPT		3.2142	2.2144	1.6308	10.4702	6.2336	4.4460
FSDT	10	1.1598	0.9443	0.7708	6.0487	2.6109	2.1091
CLPT		1.000	0.7457	0.5578	5.5710	2.3451	1.7473
FSDT	20	0.7013	0.6024	0.5180	4.2843	1.6234	1.3787
CLPT		0.5418	0.4120	0.3133	3.6574	1.3432	1.0104
FSDT	30	0.5312	0.4706	0.4170	3.2417	1.2252	1.0749
CLPT		0.3718	0.2847	0.2179	2.7376	0.9433	0.7121

5.6–3 Vibration and Buckling Analyses

The Lévy solution procedure and finite element models developed can also be used to determine the natural frequencies and critical buckling loads. Numerical results for antisymmetric cross–ply laminates are presented. In all of the examples presented here, the material properties listed in Eq. (2) were used. In the finite element analysis, a mesh of 2×2 quadratic elements was used for the FSDT, and a 4×4 mesh of four–node elements (with C^1 interpolation) was used for CLPT.

The effect of orthotropy and number of layers on the fundamental frequencies of simply supported, cross–ply, square laminates can be seen from the results presented in Table 5.6–10. The fundamental frequencies increase with an increase in degree of orthotropy and number of layers. Similar results for critical buckling loads under uniaxial compression are included in Table 5.6–11. The results of Noor [32] were obtained by applying a finite-difference scheme to the equations of the three–dimensional elasticity theory.

Table 5.6–10: Effect of degree of orthotropy of the individual layers on the dimensionless fundamental frequency of simply supported antisymmetric cross–ply square laminates: $a/h = 5, \bar{\omega} = \omega(\rho h^2/E_2)$.

Source	Layers	$\frac{E_1}{E_2} = 3$	10	20	30	40
Noor [32]		0.25031	0.27938	0.30698	0.32705	0.34250
FSDT	2	0.24834	0.27757	0.30824	0.33284	0.35333
CLPT		0.27082	0.30968	0.35422	0.39335	0.42884
Noor		0.26182	0.32478	0.37622	0.40660	0.42719
FSDT	4	0.26017	0.32898	0.38754	0.42479	0.45083
CLPT		0.28676	0.38877	0.49907	0.58900	0.66690
Noor		0.26440	0.33657	0.39359	0.42783	0.45091
FSDT	6	0.26228	0.33676	0.39771	0.43531	0.46105
CLPT		0.28966	0.40215	0.52234	0.610637	0.70359
Noor		0.26583	0.34250	0.40337	0.44011	0.46498
FSDT	10	0.26335	0.34053	0.40255	0.44023	0.46577
CLPT		0.29115	0.40888	0.53397	0.63489	0.72184

Table 5.6–11: Effect of degree of orthotropy of the individual layers on the dimensionless critical buckling loads, $\overline{N} = N_{xx}b^2/(E_2h^3)$, of simply supported (SS–1) antisymmetric cross–ply square laminates ($a/h = 10$) under uniaxial compression.

Source	Layers	$\frac{E_1}{E_2} = 3$	10	20	30	40
Noor		4.6948	6.1181	7.8196	9.3746	10.817
FSDT	2	4.7718	6.2465	8.0423	9.7347	11.353
CLPT		5.0338	6.7033	8.8158	10.891	12.957
Noor		5.1738	9.0164	13.743	17.783	21.280
FSDT	4	5.2543	9.2552	14.332	18.815	22.806
CLPT		5.5738	10.295	16.988	23.675	30.359
Noor		5.267	9.6051	15.001	19.6394	23.669
FSDT	6	5.343	9.7893	15.394	20.280	24.577
CLPT		5.674	10.960	18.502	26.042	33.582
Noor		5.3159	9.9134	15.669	20.6347	24.964
FSDT	10	5.3884	10.060	15.927	21.008	25.450
CLPT		5.725	11.300	19.277	27.254	35.232

The effect of transverse shear deformation and boundary conditions on the fundamental frequencies of two–layer and ten–layer antisymmetric cross–ply laminates ($a/h = 10$) are examined in Table 5.6–12. The first row corresponds to the analytical solution and the second row to the finite element solution. The critical buckling loads for the same laminates under uniaxial compression are presented in Table 5.6–13. In all cases, the classical plate theory overpredicts frequencies and buckling loads.

Figure 5.6–7 shows the effect of side–to–thickness ratio on the nondimensionalized fundamental frequency of cross–ply rectangular plates. Similarl results are presented for angle–ply plates in Figure 5.6–8, which shows the nondimensionalized frequency versus the lamination angle for two– and four–layer angle–ply plates ($a/b = 1, a/h = 10$).

Table 5.6–12: Effect of number of layers and transverse shear deformation on the dimensionaless frequencies $\overline{\omega} = (\omega b^2/h)(\rho/E_2)^{1/2}$ of antisymmetric cross–ply square plates $(a/h = 10)$ including various boundary conditions (Rows 1, 3, 5, and 7 correspond to anlytical results and the other rows to finite element results).

Layers	Theory	FF	FS	FC	SS	SC	CC
2	FSDT	6.881	7.215	7.741	10.473	12.610	15.152
		6.886	7.222	7.714	10.480	12.791	15.181
2	CLPT	7.267	7.636	8.228	11.154	14.223	18.543
		7.262	7.345	7.821	11.383	14.828	19.053
10	FSDT	10.900	11.079	11.862	15.779	18.044	20.471
		10.906	11.088	11.7887	15.787	18.214	20.493
10	CLPT	12.680	12.906	13.759	18.492	23.971	31.709
		12.419	11.283	11.983	18.637	23.991	31.912

Table 5.6–13: Effect of number of layers and transverse shear deformation on dimensionless critical buckling loads, $\overline{N} = N_{yy}b^2/E_2h^3$, of antisymmetric cross–ply square plates $(a/h = 10)$ including various boundary conditions (Rows 1, 3, 5, and 7 correspond to analytical results and the other rows to finite element results).

Layers	Theory	FF	FS	FC	SS	SC	CC
2	FSDT	4.851	5.351	6.166	11.353	16.437	20.067
		4.916	5.420	6.234	11.485	18.338	21.916
2	CLPT	5.425	6.003	6.968	12.957	21.116	31.280
		5.616	6.292	7.203	14.520	23.869	37.106
10	FSDT	12.092	12.524	14.358	25.450	32.614	34.837
		12.226	12.661	14.480	25.647	33.970	36.129
10	CLPT	16.426	17.023	19.389	35.232	59.288	89.770
		16.457	17.141	19.422	36.384	60.406	90.833

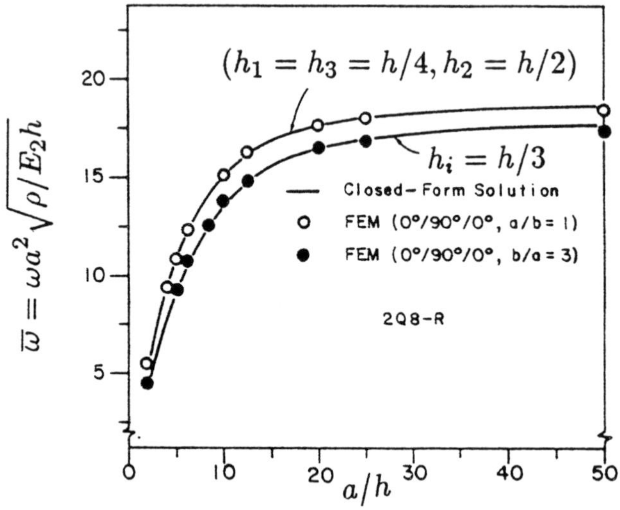

Figure 5.6–7: Effect of side–to–thickness ratio on nondimensionalized frequencies of cross–ply laminates.

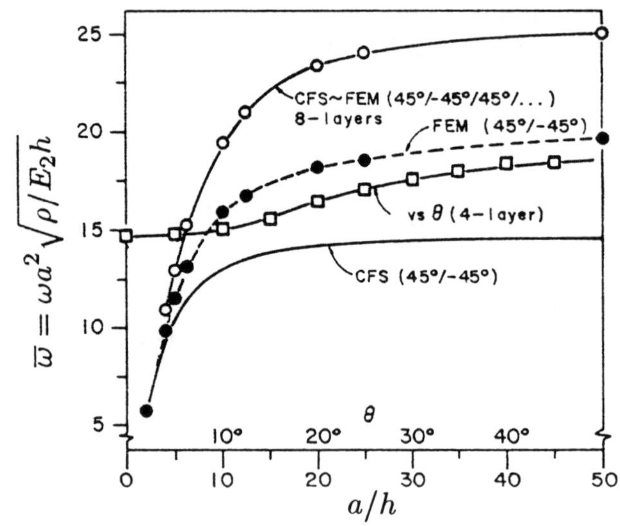

Figure 5.6–8: Effect of side–to–thickness ratio on nondimensionalized frequencies of angle–ply laminates.

The dimensionless frequencies, $\bar{\omega} \equiv \omega(a^2/h)\sqrt{(\rho/E_2)}$, of antisymmetric angle–ply laminates ($\theta°/-\theta°/\theta°/\cdots$), under various boundary conditions and with different values of θ are presented in Table 5.6–14. The critical buckling loads, $\overline{N} = N_{cr}a^2/E_2h^3$, for the same laminates under uniaxial compressive load are included in Table 5.6–15. The material properties used are those listed in Eq. (2). The results were obtained using the Lévy method with state–space approach (see Khdeir [39,43]).

Table 5.6–14: Effect of ply angle (θ) and number of layers (n) on dimensionless fundamental frequency, $\bar{\omega}$, of antisymmetric angle–play ($\theta°/-\theta°/\theta°/\ldots/-\theta°$) square plates ($a/h = 10$).

$\theta°$	n	Theory	SS	SC	CC	FF	FS	FC
	2	FSDT	12.68	13.46	14.41	6.95	8.45	8.65
30		CLPT	14.24	15.44	17.00	7.58	9.35	9.69
	10	FSDT	18.51	19.11	19.81	10.11	12.33	12.48
		CLPT	23.95	25.59	27.58	12.37	15.38	15.84
	2	FSDT	13.04	14.23	15.63	4.76	7.13	7.52
45		CLPT	14.64	16.75	19.48	5.12	7.79	8.48
	10	FSDT	19.38	20.27	21.25	6.57	10.60	10.88
		CLPT	25.47	28.91	33.32	7.89	13.03	14.17
	2	FSDT	12.68	14.52	16.57	3.33	5.87	6.70
60		CLPT	14.24	17.74	22.31	3.47	6.26	7.54
	10	FSDT	18.51	19.82	21.21	3.82	8.53	9.22
		CLPT	23.95	29.86	37.62	4.32	9.92	11.96

The analytical and finite element solutions of rectangular plates presented in this chapter for various boundary conditions serve as a reference for comparison with other approximate solutions. They can also be used to understand the behavior of laminated plates. The solutions developed herein can also be used, with the aid of a computer, to generate design data or conduct parametric studies. Such a computer software is discussed in Chapter 10 of this book.

Table 5.6–15: Effect of ply angle (θ) and number of layers (n) on dimensionless critical buckling load, $\overline{N}$, of antisymmetric angle–play ($\theta°/-\theta°/\theta°/\ldots/-\theta°$) square plates ($a/h = 10$).

$\theta°$	n	Theory	SS	SC	CC	FF	FS	FC
	2	FSDT	16.613	18.718	21.447	4.991	7.372	7.739
30		CLPT	20.543	24.158	29.269	5.822	8.857	9.520
	10	FSDT	34.931	37.242	38.920	10.423	15.520	15.899
		CLPT	58.135	66.322	77.065	15.499	23.972	25.412
	2	FSDT	17.552	19.443	19.957	2.327	5.220	5.824
45		CLPT	21.709	28.423	34.963	2.654	6.150	7.283
	10	FSDT	33.173	33.263	33.356	4.401	11.484	12.092
		CLPT	65.714	84.707	102.59	6.300	17.189	20.332
	2	FSDT	13.863	14.722	15.598	1.136	3.543	4.605
60		CLPT	19.564	23.834	29.547	1.221	3.975	5.756
	10	FSDT	23.710	23.759	23.808	1.492	7.432	8.682
		CLPT	52.945	64.103	79.619	1.889	9.977	14.501

REFERENCES FOR ADDITIONAL READING

1. Khdeir, A. A., Reddy, J. N., and Librescu, L., "Lévy Type Solutions for Symmetrically Laminated Rectangular Plates Using First–Order Shear Deformation Theory," *Journal of Applied Mechanics,* **54**, pp. 640–642, 1987.

2. Franklin, J. N., *Matrix Theory,* Prentice–Hall, Englewood Cliffs, NJ, 1968.

3. Brogan, W. L., *Modern Control Theory,* Prentice–Hall, Englewood Cliffs, NJ, 1985.

4. Reddy, J. N., *Mechanics of Laminated Composite Materials and Structures: Theory and Analysis,* John Wiley & Sons, New York, 1996.

5. Reddy, J. N., *Energy and Variational Methods in Applied Mechanics*, John Wiley & Sons, New York, 1984.

6. Reddy, J. N. and Khdeir, A. A., "Buckling and Vibration of Laminated Composite Plates Using Various Plate Theories," *AIAA Journal,* **27** (12), pp. 1808–1817, 1989.

7. Nosier, A. and Reddy, J. N., "On Vibration and Buckling of Symmetric Laminated Plates According to Shear Deformation Theories," *Acta Mechanica,***94** (3–4), pp. 123–170, 1992.

8. Reddy, J. N. and Chao, W. C., "A Comparison of Closed-Form and Finite Element Solutions of Thick Laminated Anisotropic Rectangular Plates," *Nuclear Engineering and Design*, **64**, pp. 153–167, 1981.

9. Reddy, J. N., *An Introduction to the Finite Element Method,* Second Edition, McGraw–Hill, New York, 1993.

10. Reddy, J. N., "A Penalty Plate-Bending Element for the Analysis of Laminated Anisotropic Composite Plates," *International Journal for Numerical Methods in Engineering*, **15**, pp. 1187–1206, 1980.

11. Reddy, J. N., "Analysis of Layered Composite Plates Accounting for Large Deflections and Transverse Shear Strains," in *Recent Advances in Nonlinear Computational Mechanics*, E. Hinton, D. R. J. Owen, and C. Taylor (eds.), pp. 155–202, 1982.

12. Reddy, J. N., "Geometrically Nonlinear Transient Analysis of Laminated Composite Plates," *AIAA Journal*, **21** (4), pp. 621–629, 1983.

13. Reddy, J. N., "On Refined Computational Models of Composite Laminates," *International Journal for Numerical Methods in Engineering*, **27**, pp. 361–382, 1989.

14. Hughes, T. J. R., *The Finite Element Method*, Prentice–Hall, Englewood Cliffs, NJ, 1986.

15. Bathe, K. J., *Finite Elements in Engineering Analysis*, Prentice–Hall, Englewood Cliffs, NJ, 1982.

16. Averill, R. C. and Reddy, J. N., "On the Behavior of Plate Elements Based on the First Order Shear Deformation Theory," *Engineering Computations,* **7** (1), pp. 57–74, 1990.

17. Zienkiewicz, O. C., Taylor, R. L. and Too, J. M., "Reduced Integration Technique in General Analysis of Plates and Shells," *International Journal for Numerical Methods in Engineering*, **3**, pp. 275–290, 1971.

18. Pawsey, S. F. and Clough, R. W., "Improved Numerical Integration of Thick Shell Finite Elements," *International Journal for Numerical Methods in Engineering*, **3**, pp. 575–586, 1971.

19. Hughes, T. J. R., Cohen, M. and Haroun, M., "Reduced and Selective Integration Techniques in the Finite Element Analysis of Plates," *Nuclear Engineering and Design*, **46**, pp. 203–222, 1978.

20. Malkus, D. S. and Hughes, T. J. R., "Mixed Finite Element Methods–Reduced and Selective Integration Techniques: A Unification of Concepts," *Computer Methods in Applied Mechanics and Engineering*, **15**, pp. 63–81, 1978.

21. Lee, S. W. and Pian, T. H. H., "Improvement of Plate and Shell Finite Elements by Mixed Formulations," *AIAA Journal*, **16**, pp. 29-34, 1978.

22. Spilker, R. L. and Munir, N. I., "The Hybrid-Stress Model for Thin Plates," *International Journal for Numerical Methods in Engineering*, **15**, pp. 1239–1260, 1980.

23. Park, K. C. and Stanley, G. M., "A Curved C^0 Shell Element Based on Assumed Natural-Coordinate Strains," *Journal of Applied Mechanics*, **53**, pp. 278–290, 1986.

24. Bathe, K. J., Dvorkin, E. N. and Ho, L. W., "On Discrete–Kirchhoff and Isoparametric Shell Elements for Nonlinear Analysis — An Assessment," *Computers & Structures*, **16**, pp. 89–98, 1983.

25. Zienkiewicz, O. C. and Hinton, E., "Reduced Integration, Function Smoothing and Non-Conformity in Finite Element Analysis (with special reference to thick plates)," *Journal of the Franklin Institute*, **302** (5 & 6), pp. 443–461, 1976.

26. Pugh, E. D. L., Hinton, E. and Zienkiewicz, O. C., "A

Study of Quadrilateral Plate Bending Elements with 'Reduced' Integration," *International Journal for Numerical Methods in Engineering*, **12**, pp. 1059–1079, 1978.

27. Barlow, J., "Optimal Stress Location in Finite Element Models," *International Journal for Numerical Methods in Engineering*, **10**, pp. 243–251, 1976.

28. Barlow, J., "More on Optimal Stress Points — Reduced Integration Element Distortions and Error Estimation," *International Journal for Numerical Methods in Engineering*, **28**, pp. 1486–1504, 1989.

29. Pagano, N. J., "Exact Solutions for Composite Laminates in Cylindrical Bending," *Journal of Composite Materials*, **3**, pp. 398–411, 1967.

30. Pagano, N. J., "Exact Solutions for Rectangular Bidirectional Composites and Sandwich Plates," *Journal of Composite Materials*, **4**, pp. 20–34, 1970.

31. Reddy, J. N. and Hsu, Y. S., "Effects of Shear Deformation and Anisotropy on the Thermal Bending of Layered Composite Plates," *Journal of Thermal Stresses*, **3**, pp. 475–493, 1980.

32. Noor, A. K., "Free Vibrations of Multilayered Composite Plates," *AIAA Journal*, **11**, pp. 1038–1039, 1973.

33. Noor, A. K., "Stability of Multilayered Composite Plates," *Fibre Science and Technology*, **8**(2), pp. 81–89, 1975.

34. Khdeir, A. A., Librescu, L., and Reddy, J. N., "Analytical Solution of a Refined Shear Deformation Theory for Rectangular Composite Plates," *International Journal of Solids and Structures*, **23**(10), pp. 1447–1463, 1987.

35. Khdeir, A. A. and Reddy, J. N., "Dynamic Response of Antisymmetric Angle–Ply Laminated Plates Subjected to Arbitrary Loading," *Journal of Sound and Vibration*, **126**(3), pp. 437–445, 1988.

36. Khdeir, A. A. and Librescu, L., "Analysis of Symmetric Cross–Ply Laminated Elastic Plates Using a Higher–Order Theory:

Part I–Stress and Displacement," *Composite Structures,* **9**, pp. 189–213, 1988.

37. Khdeir, A. A. and Librescu, L., "Analysis of Symmetric Cross–Ply Laminated Elastic Plates Using a Higher–Order Theory: Part II–Buckling and Free Vibration," *Composite Structures,* **9**, pp. 259–277, 1988.

38. Khdeir, A. A., "Free Vibration and Buckling of Symmetric Cross–Ply Laminated Plates by an Exact Method," *Journal of Sound and Vibration,* **126**(3), pp. 447–461, 1988.

39. Khdeir, A. A., "Free Vibration of Antisymmetric Angle–Ply Laminated Plates Including Various Boundary Conditions," *Journal of Sound and Vibration,* **122**(2), pp. 377–388, 1988.

40. Khdeir, A. A., "Free Vibration and Buckling of Unsymmetric Cross–Ply Laminated Plates Using a Refined Theory," *Journal of Sound and Vibration,* **128**(3), pp. 377–395, 1989.

41. Khdeir, A. A. and Reddy, J. N., "Exact Solutions for the Transient Response of Symmetric Cross–Ply Laminates Using a Higher–Order Plate Theory," *Composite Science and Technology,* **34**, pp. 205–224, 1989.

42. Khdeir, A. A., "An Exact Approach to the Elastic State of Stress of Shear Deformable Antisymmetric Angle–Ply Laminated Plates," *Composite Structures,* **11**, pp. 245–258, 1989.

43. Khdeir, A. A., "Comparison Between Shear Deformable and Kirchhoff Theories for Bending, Buckling and Vibration of Antisymmetric Angle–ply Laminated Plates," *Composite Structures,* **13**, pp. 159–172, 1989.

44. Khdeir, A. A., "Stability of Antisymmetrtic Angle–Ply Laminated Plates," *ASCE Journal of Engineering Mechanics,* **115**, pp. 952–962, 1989.

45. Khdeir, A. A. and Reddy, J. N., "Analytical Solutions of Refined Plate Theories of Cross–Ply Composite Laminates," *Journal of Pressure Vessel Technology,* **113**(4), pp. 570–578, 1991.

46. Khdeir, A. A. and Reddy, J. N., "Thermal Stresses and Deflections of Cross–Ply Laminated Plates Using Refined Plate Theories," *Journal of Thermal Stresses*, **14**(4), pp. 419–438, 1991.

Chapter Six

FAILURES IN COMPOSITE LAMINATES

6.1 INTRODUCTION

Composite structures are subjected to a variety of loading conditions during their service life. Under service conditions, composite structures develop matrix cracks, fiber–matrix debonds, fiber fractures, fiber pull-out, and delaminations. These effects, which cause permanent loss of stiffness and strength within the layer and hence in the laminate, are termed failures. When the failure is at the micro level, it is termed *damage*. For example, the microcracks observed within a layer constitute damage. As these microcracks grow in size and number, they coalesce and develop into debonds, resulting in a reduction of the load–carrying capacity of the laminate. In order to determine the load–carrying capacity and service life of a composite structure, it is necessary to predict the initiation and evolution of damage.

The main difficulty in mathematical modeling of damage and failures is due to the different geometric scales that are involved in the initiation and progression of damage. At the micro–level the focus is on the fiber and the matrix and fiber/matrix interface; at macro-level the focus is on the global response of the structure. The first failure always initiates at a micro–level and gradually evolves into a macro failure. However, when the objective is to predict the failure behavior of a laminate or a structure, it becomes prohibitively difficult and expensive to model the events taking place at the micro–level. Therefore, in the failure analysis of laminated composite structures, attention is often focussed on ply-level failures.

The objective of this chapter is to present a review of the literature on matrix cracking and stiffness reduction, delamination failures, and fiber pull out. A review of the failure criteria, which are used in first–ply failure determination, is also presented. For examples of various types of damage and failure studies, the reader may consult the references listed at the end of this chapter.

6.2 MATRIX CRACKS

Matrix cracks are particularly insidious since an immediate consequence is a loss of load carrying ability in a direction normal to the cracks. This manifests as a reduced stiffness in that direction. Matrix cracks by their very nature, introduce multiple stress concentration points, namely, the crack tips. These crack tips exist at the interface between plies of a laminated composite, and therefore act as potential sites of delamination, thus resulting in damage to the structure on much larger scales.

Matrix cracking is one of the first damage modes seen in angle–ply laminates under tensile loads. These matrix cracks occur in the off–axis plies of the laminate. In the absence of transverse plies, matrix cracks are seen to occur in the plies whose fiber directions make the largest angle with the direction of loading. In the 90° plies, the matrix cracks are called *transverse cracks*, since the crack faces are transverse to the loading direction. These cracks occur under monotonic tensile loading, fatigue or thermal loading. In cross–ply laminates made of glass epoxy, these cracks appear at about 0.4% applied strain and then increase in number as the strain is increased.

The problem of stiffness reduction due to the presence of matrix cracks has received considerable attention in the past and several different approaches to model stiffness reduction have been proposed. These include the shear lag model, the variational model, and the continuum damage model.

6.2–1 Shear Lag Models

Garrett and his coworkers [1] conducted extensive tension tests on cross–ply glass reinforced polyester and found that extensive cracking of the transverse ply occurred at strains much lower than the resin failure strain. The cracks showed a remarkably even spacing. Moreover, the spacing was found to decrease with increase in applied stress and increase with the thickness of the 90° plies. Garrett *et al.* started with a fundamental shear lag assumption that all load transfer from the transverse ply group to the longitudinal plies occurs at the interface and that the transverse plies are free of shear. The load transfer equation was derived using a simple force balance of the longitudinal plies. The shear stress at the interface is assumed to be proportional to the difference in the axial displacement of the longitudinal and transverse plies.

Highsmith and Reifsnider [2] found that, after sufficient loading, the off–axis plies in $[0°/90°_3]_s$, $[90°_3/0°]_s$, $[0°/90°]_s$ and $[0°/\pm 45°]_s$, glass–epoxy laminates reached a state of saturation crack spacing, where the distance between consecutive cracks in a particular ply is uniform throughout that ply. This was called the *characteristic damage state* (CDS) and is by far one of the most important observations with regard to matrix cracking.

Talug [3] formulated a finite difference solution to the stress analysis problem in the vicinity of transverse cracks. Highsmith and Reifsnider [2] observed from edge replica specimens that most of the shear deformation in any–ply was restricted to the interface region and called it the shear transfer region. A one–dimensional shear lag model was formulated based on ply equilibrium in the loading direction. The shear lag model predicted a conservative stiffness change, but was similar to the experimental results in trend.

Fukunaga *et al.* [4] used a statistical strength analysis approach including the effects of thermal residual stresses and Poisson deformation to study the failure characteristics of $[0°/90°]_s$ and $[90°/0°]_s$ laminates. Flaggs [5] made use of a mixed mode strain energy release rate criterion along with an approximate two–dimensional shear lag model including thermal loading to obtain laminate strains at the onset of transverse cracking. Laws and Dvorak [6] accounted for the existence of residual stresses in their analysis. Stiffness reduction predictions of the model in the case of E–glass epoxy laminate agreed well with experimental data at higher crack densities but violated the Hashin lower bound [7] at lower crack densities. This was attributed to a higher value of the critical strain energy release rate chosen.

Han, Hahn and Croman [8] analyzed transverse cracks in cross-ply laminates based on the concept of through–the–thickness inherent flaw and principle of energy balance. The stress predictions of the analysis were compared with a plane strain finite element model of $[0°/90°_3]_s$, T300/934 graphite epoxy composite laminate. The crack opening displacement computed from the model agreed well with the finite element results.

Lim and Hong [9] obtained the stress state in the vicinity of a transverse matrix crack by first assuming a displacement field for in-plane displacements in terms of unknown functions that are solved as solutions to the in-plane shear lag equations. Kim and Hong [10] worked on a progressive failure model based on concepts of fiber bundle failure and matrix cracking. They extended the analysis of Lim and Hong separately for tensile and shear loads. Zhang, Fan and Zoutis [11] developed a two–dimensional shear lag model in which a linear variation of the transverse shear stress across the thickness of the transverse plies and partial linear variation across the thickness of the constraining group of plies in $[\pm\theta^\circ_m/90^\circ_n]_s$ laminates was considered.

Lee and Hong [12] worked on a refined two–dimensional model by extending the concept of interlaminar shear layer to an interface layer. Guild *et al.* [13] examined the mechanics of transverse cracking using a three–dimensional finite element model. Predictions of the analysis agreed with the parabolic shear lag model.

While a shear lag model provides a basic understanding of the load transfer process in laminates with matrix cracks in the transverse plies, there are many shortcomings of this model. Most shear lag models assume that the load transfer between the cracked group of plies and the constraining group of plies occurs at the interface between these groups. Edge replica studies have indeed supported this idea. However, all the shear stress is assumed to be restricted to the interface alone thereby introducing a shear discontinuity into the model. This assumes that the transverse ply is shear stress free through its thickness, and therefore shear strain free due to the orthotropy. This amounts to shear stiffening the plies, and one therefore expects the shear lag model to predict a higher stiffness in general, which they do. Another common assumption is that the in-plane displacements are constant through the thickness of each group of plies and that the axial stress perturbation in both the longitudinal and transverse plies is constant in the direction of the thickness coordinate. This precludes all crack tip effects on the axial stress and therefore the existence of a disturbance zone in the constraining plies. Clearly, if the crack is not able to propagate through the thickness, beyond the transverse ply, the resistance to growth and hence the additional axial stress developed in the longitudinal ply must be greater at and near the immediate vicinity of the crack tip than farther away from it. This implies that it is necessary to model axial stress perturbations in the plies so as to capture the thickness coordinate dependence.

Moreover, as is stated by Nairn [14], the shear lag model requires that the stiffness of the load bearing region be significantly larger than that of the shear transfer region, which is very thin. In cross–ply laminates, the axial stiffness of the longitudinal plies is not very much greater than the stiffness of this shear transfer region, and, therefore, the shear lag requirement breaks down. Another simplifying assumption is that the crack interactions are neglected. It is expected that crack interaction will increase with crack density. Thus the shear lag approach is limited in accuracy in so far as this interaction has been neglected. Moreover, most shear lag models are heuristic to the extent of requiring shear lag parameters that are characteristics of the shear transfer region. These typically depend on the thickness and elastic constants of the shear transfer region. In most cases, the matrix properties are assigned to the shear layer since it is assumed that this layer is a resin rich layer. This leaves the thickness of the interlaminar shear transfer region adjustable.

Also, the equilibrium equations of the shear lag model are satisfied only in an approximate sense. Yet another drawback is that the analysis becomes rather cumbersome in the case of laminates with layup schemes like $[90^\circ_m/0^\circ_n]_s$ and in laminates with transverse plies distributed arbitrarily through the thickness since, in these layup schemes, the transverse cracks are found to occur in a staggered manner. It is rather difficult to adopt this model to study effects of transverse cracking under biaxial loading conditions.

6.2–2 Variational Models

Another approach to the stiffness reduction problem in composite laminates with transverse matrix cracks in the off–axis plies is the "variational approach", based on energy minimization principles. While being a simpler approach, it admits the relaxation of some of the assumptions and eliminates the "adjustability" requirement of the shear lag model.

Hashin [7] developed a variational model to determine an approximate stress state in the vicinity of a transverse crack. The analysis entailed the assumption, as is common to most shear lag theories, that the axial stress perturbation within each group of plies is a function of the axial coordinate and is independent of the thickness coordinate. The principle of minimum complementary energy was invoked to obtain the stress perturbations separately for the cases of uniaxial tension and in-plane shear. The stiffness reduction of $[0°/90°]_s$ was found to agree very well with the data of Highsmith and Reifsnider [2]. The analysis gives a strict lower bound for the stiffnesses.

Herakovich *et al.* [15] developed an approximate analytical model that accounted for uniaxial tension or in–plane shear or thermal loading. A mixed boundary value problem was solved for an RVE (representative volume element), and the axial and normal displacement field for different ply groups were obtained as second order Legendre expansions in the ply thickness coordinate with the constants of the expansion being functions of the axial coordinate. The average axial stress through the thickness of the laminate and the average strain were then calculated to arrive at the effective axial stiffness and the Poisson's ratio for the laminate. A finite element analysis was also carried out based on a displacement formulation. A comparison with the finite element stiffness predictions was made for $[0^\circ_3/90^\circ]_s$, $[0^\circ_2/90^\circ_2]_s$ and $[0^\circ/90^\circ_3]_s$ graphite epoxy laminates.

Building on the approach of Hashin on the stiffness reduction problem in cross-ply laminates, Nairn [14] analyzed $[0^\circ_m/90^\circ_n]$ and $[90^\circ_m/0^\circ_n]$ laminates by including the effect of residual thermal stresses. Varna and Berglund [16,17] extended the thermoelastic analysis of the $[0^\circ_m/90^\circ_n]_s$ laminates by taking into account a nonlinear stress distribution in the thickness direction of the constraining 0° plies.

McCartney [18] analyzed the stress transfer between the constraining ply and the transverse ply in a $[0°/90°]_s$ cross-ply laminate containing matrix cracks. The thermal residual stresses arising from mismatch in thermal expansion of the longitudinal and transverse plies were accounted for. A minimum complementary energy approach was utilized thus satisfying the equilibrium equations, the interface conditions, and the stress boundary conditions exactly. Analysis was carried out for both plane strain and generalized plane strain conditions. Theoretical predictions for glass fiber epoxy and carbon fiber epoxy laminates, of the Young's modulus and Poisson's ratios, compared well with experiments. The plane strain stiffness reduction was predicted to be higher than that for generalized plane strain. Gamby and Rebiére [19] assumed a Fourier series representation of the transverse shear stresses in the longitudinal and transverse ply groups.

6.2–3 Continuum Damage Models

Talreja [20] considered the problem of stiffness reduction in composite laminates with matrix cracking and delamination by introducing a damage vector concept to evaluate internal state variables that quantify damage. Results of change in longitudinal stiffness with cycles of loading in the case of $[0°/90°_3]_s$ glass epoxy laminate under tensile fatigue load agreed reasonably well with experiments.

Allen *et al.* [21,22] proposed a continuum damage–mechanics based concept wherein damage in laminates was characterized by a set of internal state variables (ISV). The internal state variables represented locally averaged measures of the state of damage and were typically second order tensorial quantities. The ISV for matrix cracks was related to the energy release rate due to cracking, using the concept of linear elastic fracture mechanics. Predictions of the theory compared well with experiments.

Lee *et al.* [23] proposed an upper bound based on the internal state variable approach. The displacement field in the presence of cracks was assumed in terms of an infinite series in trigonometric functions in order to evaluate the internal state variables. Tay and Lim [24] used a constitutive model of the damage state proposed by Allen *et al.* [21,22]. They performed a series of parametric finite element analyses to determine the effects of relative ply thickness of the longitudinal and the transverse plies, crack density, and crack opening displacement on the vector of internal state variables.

6.2–4 Elasticity Models

Nuismer and Tan [96] developed an approximate elasticity theory solution for the stress strain relations of a cracked lamina. The effective constitutive coefficients were solved in terms of average stresses and average strains. The predicted reduction of axial and shear stiffness was found to agree with finite element results. Wu and Chudnovsky [25] analyzed the relation between effective elastic properties of laminates with matrix cracks by a beam model for the two-dimensional case where the length of the representative element was much larger than its thickness.

Another approach to the stiffness reduction problem due to Gudmundson and Östlund [26], applicable to thin and thick laminates, is based on the concept of change in elastic energy due to the appearance of cracks in the plies. The cracks are assumed to be noninteracting and "dilute" in that the crack density was low. Stiffness reduction results for $[0°/90°]_s$ glass epoxy laminate compared well with the experimental results of Highsmith and Reifsnider [2]. Yuan and Selek [27] presented an elasticity model to study the stiffness reduction in $[0^\circ_n/90^\circ_m]_s$ and $[\theta^\circ_n/90^\circ_m]_s$ laminates. The crack tip stresses and the displacements were solved in terms of Lekhnitskii's complex potentials and the method of eigenfunction expansion. The axial stiffness reduction was found to agree well with experiments and the predictions of the shear lag theory.

6.2–5 Self–Consistent Model and Other Models

Self–consistent models are used to study the effect of inhomogeneities in homogeneous solids. Cracks are modeled as inhomogeneities and the actual configuration is approached in the limiting case of vanishing stiffness and the inhomogeneities being shrunk to surfaces (see Kachanov [28]). An example of the effective field concept is the Mori–Tanaka method.

Dvorak *et al.* [29] used a self–consistent scheme to evaluate the overall stiffness and thermal expansion coefficients and the stress and strain averages for a composite lamina containing transverse cracks with a known density and subjected to mechanical and thermal loads. They modeled the laminate with cracks as a two–phase model by treating cracks as limiting ellipsoids. Aboudi [30] presented a method for the determination of the effective elastic moduli of elastic solids containing a doubly periodic rectangular array of cracks.

Sumarac and Krajcinovic [31] developed a self-consistent process model for a perfectly elastic solid weakened by an ensemble of microcracks. The study was limited to a weak interaction between cracks. Bai [32] formulated the problem in terms of a singular integral equation that was solved numerically, and the results agreed well with the Hashin lower bound solution. Other supplementary sources consulted but not discussed in detail are Aveston [33], Brillaud and El Mahi [34], Caslini *et al.* [35], Ogin *et al.* [36], Wang *et al.* [37], and Praveen and Reddy [38,39].

6.3 DELAMINATIONS

The effect of delamination on stiffness and strength is not as well understood as the effect of matrix cracking. Most of the effort in the area of delamination is directed towards predicting the initiation and growth of delaminations. O'Brien [40,41] analyzed local delaminations and studied their influence on tensile stiffness of composite laminates. He also studied [42] the effect of delamination on the strength of unnotched quasi–isotropic graphite/epoxy laminates.

The edge delamination tension (EDT) test is used to measure the mixed–mode interlaminar fracture toughness of composite laminates [43–49]. In the EDT test, a composite laminate strip is loaded in tension to produce high interlaminar tensile and shearing stresses along the free edges thus causing delamination. Delamination onset and growth are typically correlated with the total energy release rate G, or with some combination of the individual components (G_I, G_{II}, and G_{III}) of G. O'Brien [40] introduced a simple rule–of–mixtures formula based on classical laminate theory for computing the total energy release rate at delamination onset. To determine the individual components of G, quasi–3D finite element analyses were used in conjunction with the virtual crack closure method [45–50].

Typically it is assumed that an existing edge delamination in an EDT specimen will open over the entire delaminated area during loading; however for certain laminates, the delamination actually closes over a portion of the delaminated area. In References 43–49, the effect of possible closure of the delamination during loading was not considered. When this boundary condition is not accounted for, numerical analyses of the EDT test specimen may predict a physically inadmissible overlapping of the interlaminar crack faces. To investigate the effect of preventing crack surface overlapping on the computed energy release rate G and its individual components, the EDT test specimen was modeled by Robbins and Reddy [51] (also see Barbero and Reddy [52]) using the layer–wise displacement–based, finite element model [53]. The delamination problem analyzed consisted of an edge delamination tensile specimen $(0/\pm 35/90)_s$ with existing edge delaminations at the $-35/90$ and $90/-35$ interfaces. For this particular laminate, the standard linear finite element analysis predicts a physically inadmissible overlapping of the delaminated surfaces. An iterative technique for solving the frictionless contact problem was used to prevent surface overlap, thus resulting in a slight decrease in the total energy release rate G; however, the individual components (G_I and G_{II}) of G experienced significant changes in magnitude.

The double cantilever beam (DCB) specimen is widely used for characterizing the mode I fracture toughness of laminated composite materials and adhesively bonded joints [54–59]. Laminated composite DCB specimens are typically fabricated with an insert embedded at the midplane of one end of the beam to provide an initial straight delamination front. To avoid the effects of fiber bridging, Martin [56] stressed the importance of determining the critical energy release rate during the initial delamination propagation from the insert, and investigated the influence of insert thickness on the critical energy release rate. Davidson and Schapery [58] pointed out that, for many laminate lay-ups, the delamination front attains significant curvature during propagation and further suggested that the delamination front curvature may be caused by the tendency for the bent arms of the specimen to exhibit anticlastic curvature. Crews *et al.* [59] investigated the effect of anticlastic curvature on the variation of

G along the initial straight delamination starter using a three–dimensional finite element model. They noted that amount of anticlastic curvature (and thus the amount of variation of G along the delamination front) is related to the quantity $\nu_{xy}E_y/E_x$, which is typically largest for a $((\pm 45)n)_s$ laminate. When modeling this particular laminate, they observed that the delaminated surfaces would overlap, in the finite element model, near the intersection of the straight delamination front and the free edge. To investigate the effect of preventing crack surface overlapping on the computed energy release rate G, the $((\pm 45)n)_s DCB$ specimen was modeled by Robbins and Reddy [51] using the layerwise finite element model. The delamination problem consisted of a homogenized $((\pm 45)n)_s$ double cantilever beam specimen. A standard linear finite element analysis for this configuration predicts an overlapping of the delaminated surfaces near the intersection of the delamination front and the free edge. When the no–penetration boundary conditions were accounted for, the G_I distribution experienced significant change in the vicinity of the original overlap.

6.4 FIBER PULL/PUSH OUT

It has been recognized that in brittle matrix composites post debonding friction and fiber pull–out may be important mechanisms of energy consumption during failure [60–63]. According to Suzuki *et al.* [62] the fiber pull–out resistance may be a thousand times more than that of monolithic ceramics. A number of experimental and theoretical studies have been conducted in recent times, some of which are reviewed in Reference 64.

Though a single fiber pull–out is far removed from the realistic fracture mechanism, it has been used as a good source of study for understanding the fiber pull–out mechanism. Khanna [65] has successfully used single fiber pull–out tests to explain certain observations that have been made on the overall failure of composite materials. Also, single fiber pull–out tests have been used traditionally to understand the fiber matrix interface strength in composite materials [66–68].

Fiber pull–out has been studied theoretically by a number of investigators [69–73]. Hutchinson and Jensen [69] have considered debonding and frictional sliding, and they constructed approximate closed–form solutions based on the solutions of the Lamé problem. They also constructed a numerical solution and considered two cases, one with a constant friction and the other with Coulomb friction along the interface. Aboudi [70] performed a micromechanical analysis of the problem incorporating debonding by a critical shear stress model. Friction was accounted through a Coulomb friction law. Alternatively, Gao *et al.* [71] used the fracture mechanics approach to study debonding and determined the work of friction following debonding. They used the plane strain Lamé solutions to study the fiber/matrix interaction. Dollar and Steif [72] used a continuous distribution of dislocations to model Coulomb friction. This technique results in a singular integral equation, which is solved to determine the load transfer. Their major conclusion is that the constant shear stress approximation results in substantially inaccurate results. Fruend [73] studied the problem of a circular fiber slipping through an unbounded elastic material. This study brings out the effect of the stiffness ratio of the fiber and the matrix on the stress distribution. Again, an integral equation has been solved numerically to obtain the results.

Povirk and Needleman [74] simulated the fiber pull–out problem using a rate dependent internal variable friction constitutive equation. The fiber was considered to be rigid and the analysis was essentially concerned with the effect of the friction law in explaining the stick–slip instability during pull out. Ballarami *et al.* [75] modeled the frictional slip with the assumption that both the fiber and the matrix are made up of the same material. Marotzke [76] studied the pull–out test using finite element analysis without considering the frictional slip. Faber *et al.* [77] conducted preliminary finite element analysis of fiber push out problems for a stiff fiber in a soft matrix.

The more important problem that has not attracted the attention of the researchers is the stress distribution in the presence of a broken fiber. As observed by Khanna [65], most fiber fracture takes place not exactly at the crack plane, but away from it. Hence as the fiber is pulled out, the broken part of the fiber remains in the matrix. This broken fiber may have a "plug" effect and alter the axial and radial stresses near the free end of the pulled fiber. Another problem of interest is to determine the effect of an interfacial coating on the shear and axial stress distributions in the pulled fiber. The stress transfer during pull out with an elastic interface has been analyzed only to a limited extent (see [78]). An understanding of this phenomenon is important to design proper interfacial coatings. Krishna Kumar and Reddy [60] analyzed the three fiber pull out problems stated above for stress distributions in the fiber during fiber pull–out. The finite element model based on the perturbed Lagrangian approach and Coulomb's friction law was used. A linear elastic model with contact algorithm was used to simulate the pull–out. Their results indicate that a "pinching" effect near the free end of a fiber affects the local stresses, especially for small values of the stiffness of the matrix. Broken fiber (embedded in the matrix) does not seem to affect the over all stress distribution. The effect of coatings on the stress distribution depends on the ratios of the stiffness and the thickness of the coating. Thicker and stiffer coatings increase the overall modulus and hence produce higher stresses for the same displacement. The coating thickness has a greater influence than the coating stiffness on the shear stress distribution.

6.5 FAILURE CRITERIA

6.5–1 General Comments

Once the mechanical properties of a ply are known, the initial failure of the ply, known as the *first–ply failure* [79–81], within a laminated structure can be predicted by applying an appropriate failure criterion. The subsequent failure prediction requires a knowledge of damage modes and damage accumulation and their effect on stiffnesses. There are many proposed theories to predict the onset of failures and their progression (see [82–85]). The failure criteria consist of parameters that must be experimentally determined. Often, these parameters are difficult to determine with certainity.

All except the maximum strain criterion are based on the stress state in a lamina. If experimental data are available, a strain–based criterion is more accurate since it is directly measured. The stress–based data are difficult to generate, due to the statistical variations in obtaining stresses from strains. The major problem is that the constitutive matrix is not known precisely at the strain level. Since damage accumulation is cumulative, and failure is usually progressive rather than catastrophic and instantaneous, error introduced through assumed material stiffnesses can grow with local steps. Hence, calculation of stresses from strains of a laminate requires further study.

It should be noted that the failure criterion is used only to check whether allowables are exceeded. In the case of delaminations, the failure criterion can be used to determine the onset of delaminations. A separate criterion for the growth of delamination is required. Often such a criterion is expressed in terms of the strain energy release rates, using fracture mechanics concepts. The fracture mechanics laws to study crack growth are primarily developed for homogeneous isotropic material systems. Thus the fracture mechanics concepts without any homogenization do not hold for composite material systems. References 86–94, in addition to those cited earlier, provide models for failures and their determination in various laminated structures.

Failure criteria for composite materials can be classified into two groups, namely: (i) independent failure criteria, and (ii) polynomial failure criteria. These are reviewed below.

6.5–2 Independent Failure Criteria

The maximum stress and maximum strain criteria belong to this category. In the maximum stress criterion, failure is assumed to occur if any one of the following conditions are satisfied:

$$\begin{aligned} &\sigma_1 \geq X_T\ ,\ \sigma_4 \geq R \\ &\sigma_2 \geq Y_T\ ,\ \sigma_5 \geq S \\ &\sigma_3 \geq Z_T\ ,\ \sigma_6 \geq T \end{aligned} \qquad (6.5-1)$$

where $(\sigma_1, \sigma_2, \sigma_3)$ are the normal stress components, $(\sigma_4, \sigma_5, \sigma_6)$ are shear stress components, (X_T, Y_T, Z_T) are the lamina normal strengths in tension (T) along the (1,2,3) directions, and (R, S, T) are the shear strengths in the (23, 13, 12) planes, respectively. When $(\sigma_1, \sigma_2, \sigma_3)$ are of a compressive nature, they should be compared with (X_C, Y_C, Z_C), which are normal strengths in compression (C) along the (1,2,3) directions, respectively.

In the maximum strain criterion, failure is assumed to occur if any of the following conditions are satisfied.

$$\begin{aligned} \varepsilon_1 &\geq X_{\varepsilon T} \ , \ \varepsilon_4 \geq R_\varepsilon \\ \varepsilon_2 &\geq Y_{\varepsilon T} \ , \ \varepsilon_5 \geq S_\varepsilon \\ \varepsilon_3 &\geq Z_{\varepsilon T} \ , \ \varepsilon_6 \geq T_\varepsilon \end{aligned} \qquad (6.5-2)$$

where (ε_1, ε_2, ε_3) are the normal strain components, (ε_4, ε_5, ε_6) are shear strain components, ($X_{\varepsilon T}$, $Y_{\varepsilon T}$, $Z_{\varepsilon T}$) are the lamina normal strain strengths in tension along the (1,2,3) directions, and (R_ε, S_ε, T_ε) are the shear strain strengths in the (23, 13, 12) planes, respectively. When (ε_1, ε_2, ε_3) are negative, they should be compared with ($X_{\varepsilon C}$, $Y_{\varepsilon C}$, $Z_{\varepsilon C}$), which are normal strain strengths in compression along the (1,2,3) directions.

6.5–3 Polynomial Failure Criteria

The most general polynomial failure criterion for composite materials is the tensor polynomial criterion proposed by Tsai [82]. All other polynomial failure criteria are degenerate cases of this criterion. In index notation, the tensor polynomial failure criterion is expressed as:

$$F_i\sigma_i + F_{ij}\sigma_i\sigma_j + F_{ijk}\sigma_i\sigma_j\sigma_k + \ldots \geq 1 \ ,$$

or, in more explicit form,

$$\begin{aligned} &F_1\sigma_1 + F_2\sigma_2 + F_3\sigma_3 + 2F_{12}\sigma_1\sigma_2 + 2F_{13}\sigma_1\sigma_3 + 2F_{23}\sigma_2\sigma_3 + F_{11}\sigma_1^2 \\ &+2F_{22}\sigma_2^2 + F_{33}\sigma_3^2 + F_{44}\sigma_4^2 + F_{55}\sigma_5^2 + F_{66}\sigma_6^2 + \ldots \geq 1 \end{aligned} \qquad (6.5-3)$$

Particular cases of the above criterion are discussed below.

Maximum stress (polynomial) criterion

The maximum stress criterion can also be expressed as a tensor polynomial criterion as:

$$(\sigma_1 - X_T)(\sigma_1 + X_C)(\sigma_2 - Y_T)(\sigma_2 + Y_C)(\sigma_3 - Z_T)(\sigma_3 + Z_C)$$

$$(\sigma_4 - R)(\sigma_4 + R)(\sigma_5 - S)(\sigma_5 + S)(\sigma_6 - T)(\sigma_6 + T) = 0 \quad (6.5-4)$$

Comparing Eq. (4) with Eq. (3) and ignoring the higher order terms in both the equations, we obtain

$$\begin{aligned}
F_1 &= \frac{1}{X_T} - \frac{1}{X_C}\,, \quad F_2 = \frac{1}{Y_T} - \frac{1}{Y_C}\,, \quad F_3 = \frac{1}{Z_T} - \frac{1}{Z_C} \\
F_{11} &= \frac{1}{X_T X_C}\,, \quad F_{22} = \frac{1}{Y_T Y_C}\,, \quad F_{33} = \frac{1}{Z_T Z_C} \\
F_{44} &= \frac{1}{R^2}\,, \quad F_{55} = \frac{1}{S^2}\,, \quad F_{66} = \frac{1}{T^2} \\
F_{12} &= \frac{F_1 F_2}{2}\,, \quad F_{13} = \frac{F_1 F_3}{2}\,, \quad F_{23} = -\frac{F_2 F_3}{2} \qquad (6.5-5)
\end{aligned}$$

The remaining strength constants are zeros.

Maximum strain (polynomial) criterion

The maximum strain criterion can be expressed as a tensor polynomial criterion:

$$(\varepsilon_1 - X_{\varepsilon T})(\varepsilon_1 + X_{\varepsilon C})(\varepsilon_2 - Y_{\varepsilon T})(\varepsilon_2 + Y_{\varepsilon C})(\varepsilon_3 - Z_{\varepsilon T})(\varepsilon_3 + Z_{\varepsilon C})\times$$

$$(\varepsilon_4 - R_\varepsilon)(\varepsilon_4 + R_\varepsilon)(\varepsilon_5 - S_\varepsilon)(\varepsilon_5 + S_\varepsilon)(\varepsilon_6 - T_\varepsilon)(\varepsilon_6 + T_\varepsilon) = 0 \quad (6.5-6)$$

Using the compliance matrix and the constitutive relation, we can express Eq. (6) in the form of equation (3) as:

$$F_1 = F_1^A + \frac{S_{12}}{S_{22}} F_3^A + \frac{S_{13}}{S_{33}} F_3^A$$

$$F_2 = \frac{S_{12}}{S_{11}} F_1^A + F_2^A + \frac{S_{23}}{S_{33}} F_3^A$$

$$F_1 = \frac{S_{13}}{S_{11}} F_1^A + \frac{S_{23}}{S_{22}} F_2^A + F_3^A$$

$$F_{11} = S_x + \left(\frac{S_{12}}{S_{22}}\right)^2 S_y + \left(\frac{S_{13}}{S_{33}}\right)^2 S_z - \frac{S_{13}}{S_{33}} G_{13}^A - \frac{S_{12}}{S_{22}} G_{12}^A - \frac{S_{12}S_{13}}{S_{22}S_{23}} G_{23}^A$$

$$F_{22} = S_y + \left(\frac{S_{12}}{S_{11}}\right)^2 S_x + \left(\frac{S_{23}}{S_{33}}\right)^2 S_z - \frac{S_{12}}{S_{11}} G_{12}^A - \frac{S_{23}}{S_{33}} G_{23}^A - \frac{S_{12}S_{23}}{S_{11}S_{33}} G_{13}^A$$

$$F_{33} = S_z + \left(\frac{S_{13}}{S_{22}}\right)^2 S_x + \left(\frac{S_{23}}{S_{22}}\right)^2 S_y - \frac{S_{13}}{S_{11}} G_{13}^A - \frac{S_{23}}{S_{22}} G_{23}^A - \frac{S_{13}S_{23}}{S_{11}S_{22}} G_{12}^A$$

$$S_x = \frac{1}{X_T X_C}\ , \quad S_y = \frac{1}{Y_T Y_C}\ , \quad S_z = \frac{1}{Z_T Z_C}$$

$$F_{44} = \frac{1}{R^2}\ , \quad F_{55} = \frac{1}{S^2}\ , \quad F_{66} = \frac{1}{T^2}\ , \quad G_{ij}^A \equiv F_i^A F_j^A$$

$$F_{12} = \frac{S_{12}}{S_{11}} \frac{1}{X_T X_C} + \frac{S_{12}}{S_{22}} \frac{1}{Y_T Y_C} + \frac{S_{13}S_{23}}{S_{33}^2} \frac{1}{Z_T Z_C} - \frac{1}{2}\left(\frac{S_{12}^2}{S_{11}S_{22}} + 1\right) G_{12}^A - \frac{1}{2}\left(\frac{S_{13}S_{12}}{S_{11}S_{33}}\right) G_{13}^A - \frac{1}{2}\left(\frac{S_{12}S_{23}}{S_{22}S_{33}} + \frac{S_{13}}{S_{33}}\right) G_{23}^A$$

$$F_{13} = \frac{S_{13}}{S_{11}} \frac{1}{X_T X_C} + \frac{S_{13}}{S_{33}} \frac{1}{Z_T Z_C} + \frac{S_{12}S_{23}}{S_{22}^2} \frac{1}{Y_T Y_C} - \frac{1}{2}\left(\frac{S_{13}^2}{S_{11}S_{33}} + 1\right) G_{12}^A - \frac{1}{2}\left(\frac{S_{12}S_{13}}{S_{11}S_{22}} + \frac{S_{23}}{S_{22}}\right) G_{12}^A - \frac{1}{2}\left(\frac{S_{13}S_{23}}{S_{22}S_{33}} + \frac{S_{12}}{S_{22}}\right) G_{23}^A$$

$$F_{23} = \frac{S_{23}}{S_{22}} \frac{1}{Y_T Y_C} + \frac{S_{23}}{S_{33}} \frac{1}{Z_T Z_C} + \frac{S_{12}S_{13}}{S_1^2} \frac{1}{X_T X_C} - \frac{1}{2}\left(\frac{S_{23}^2}{S_{22}S_{33}} + 1\right) G_{12}^A - \frac{1}{2}\left(\frac{S_{12}S_{23}}{S_{11}S_{22}} + \frac{S_{13}}{S_{11}}\right) G_{12}^A - \frac{1}{2}\left(\frac{S_{23}S_{13}}{S_{11}S_{33}} + \frac{S_{12}}{S_{11}}\right) G_{23}^A \quad (6.5-7)$$

In the above expression S_{11}, S_{12}, etc. are the components of the compliance matrix; X_T, Y_T, Z_T are the tensile strengths in the 1,2,3 directions; X_C, Y_C, Z_C are the compressive strengths in the 1,2,3 directions; R, S, T are the shear strengths in the 23, 13, 12 planes, respectively; and F_1^A, F_2^A, F_3^A are the expressions given for F_1, F_2, F_3 in the maximum stress criterion. All other strength tensor terms are zero.

Tsai–Hill's criterion

Hill's criterion can be expressed as

$$\left(\frac{\sigma_1}{X}\right)^2 + \left(\frac{\sigma_2}{Y}\right)^2 + \left(\frac{\sigma_3}{Z}\right)^2 - \left(\frac{1}{X^2} + \frac{1}{Y^2} - \frac{1}{Z^2}\right)\sigma_1\sigma_2$$
$$- \left(\frac{1}{Y^2} + \frac{1}{Z^2} - \frac{1}{X^2}\right)\sigma_2\sigma_3 - \left(\frac{1}{Z^2} + \frac{1}{X^2} - \frac{1}{Y^2}\right)\sigma_1\sigma_3$$
$$+ \left(\frac{\sigma_4}{R}\right)^2 + \left(\frac{\sigma_5}{S}\right)^2 + \left(\frac{\sigma_6}{T}\right)^2 \geq 1 \qquad (6.5-8)$$

Hoffman's criterion

Hoffman's criterion is expressed as

$$\frac{1}{2}\left(S_y + S_z - S_x\right)\left(\sigma_2 - \sigma_3\right)^2 + \frac{1}{2}\left(S_z + S_x - S_y\right)\left(\sigma_3 - \sigma_1\right)^2$$
$$+ \frac{1}{2}\left(S_x + S_y - S_z\right)\left(\sigma_1 - \sigma_2\right)^2 + \left(\frac{1}{X_T} - \frac{1}{X_C}\right)\sigma_1 + \left(\frac{1}{Y_T} - \frac{1}{Y_C}\right)\sigma_2$$
$$- \left(\frac{1}{Z_T} - \frac{1}{Z_C}\right)\sigma_3 + \left(\frac{\sigma_4}{R}\right)^2 + \left(\frac{\sigma_5}{S}\right)^2 + \left(\frac{\sigma_6}{T}\right)^2 \geq 1 \qquad (6.5-9)$$

where

$$S_x = \frac{1}{X_T X_C}\ ;\ S_y = \frac{1}{Y_T Y_C}\ ;\ S_z = \frac{1}{Z_T Z_C}$$

Tsai–Wu's criterion

The Tsai–Wu criterion is given by

$$F_i\sigma_i + F_{ij}\sigma_i\sigma_j \geq 1 \qquad (6.5-10)$$

where

$$\begin{aligned}
F_1 &= \frac{1}{X_T} - \frac{1}{X_C}, \quad F_2 = \frac{1}{Y_T} - \frac{1}{Y_C}, \quad F_3 = \frac{1}{Z_T} - \frac{1}{Z_C} \\
F_{11} &= \frac{1}{X_T X_C}, \quad F_{22} = \frac{1}{Y_T Y_C}, \quad F_{33} = \frac{1}{Z_T Z_C} \\
F_{44} &= \frac{1}{R^2}, \quad F_{55} = \frac{1}{S^2}, \quad F_{66} = \frac{1}{T^2} \\
F_{12} &= -\frac{1}{2}\frac{1}{\sqrt{X_T X_C Y_T Y_C}} \\
F_{13} &= -\frac{1}{2}\frac{1}{\sqrt{X_T X_C Z_T Z_C}} \\
F_{23} &= -\frac{1}{2}\frac{1}{\sqrt{Y_T Y_C Z_T Z_C}}
\end{aligned} \qquad (6.5-11)$$

It should be noted that in Hill's criterion the linear terms in σ_i do not appear; therefore, the values of F_1, F_2 and F_3 are zero. The values of X, Y, Z are taken as either X_T, Y_T, Z_T or as X_C, Y_C, Z_C depending upon the sign of σ_1, σ_2 and σ_3, respectively. Strength tensors for this criterion are ($F_i = 0$):

$$\begin{aligned}
F_{11} &= \frac{1}{X^2}, \quad F_{22} = \frac{1}{Y^2}, \quad F_{33} = \frac{1}{Z^2} \\
F_{44} &= \frac{1}{R^2}, \quad F_{55} = \frac{1}{S^2}, \quad F_{66} = \frac{1}{T^2} \\
F_{12} &= -\frac{1}{2}\left(\frac{1}{X^2} + \frac{1}{Y^2} - \frac{1}{Z^2}\right) \\
F_{13} &= -\frac{1}{2}\left(\frac{1}{Z^2} + \frac{1}{X^2} - \frac{1}{Y^2}\right) \\
F_{23} &= -\frac{1}{2}\left(\frac{1}{Y^2} + \frac{1}{Z^2} - \frac{1}{X^2}\right)
\end{aligned} \qquad (6.5-12)$$

All other strength tensor terms are zero.

Hoffman's criterion is a special case of the tensor polynomial criterion for the following choice of the parameters F_i and F_{ij}:

$$
\begin{aligned}
F_1 &= \frac{1}{X_T} - \frac{1}{X_C} \, , \; F_2 = \frac{1}{Y_T} - \frac{1}{Y_C} \, , \; F_3 = \frac{1}{Z_T} - \frac{1}{Z_C} \\
F_{11} &= \frac{1}{X_T X_C} \, , \; F_{22} = \frac{1}{Y_T Y_C} \, , \; F_{33} = \frac{1}{Z_T Z_C} \\
F_{44} &= \frac{1}{R^2} \, , \; F_{55} = \frac{1}{S^2} \, , \; F_{66} = \frac{1}{T^2} \\
F_{12} &= -\frac{1}{2}\left(\frac{1}{X_T X_C} + \frac{1}{Y_T Y_C} - \frac{1}{Z_T Z_C}\right) \\
F_{13} &= -\frac{1}{2}\left(\frac{1}{X_T X_C} + \frac{1}{Z_T Z_C} - \frac{1}{Y_T Y_C}\right) \\
F_{23} &= -\frac{1}{2}\left(\frac{1}{Z_T Z_C} + \frac{1}{Y_T Y_C} - \frac{1}{X_T X_C}\right) \qquad (6.5-13)
\end{aligned}
$$

Other strength tensor terms are zero.

Hashin's criteria

The criteria reviewed above were originally developed to predict initial failures, without regard to the mode of failure. However, in recent years these criteria have been also used to identify damage modes (see [81]). Hashin [84] suggested a set of failure criteria that have the ability to predict modes of failures. These are reviewed here.

Fiber failure

Tensile:

$$\left(\frac{\sigma_1}{X_T}\right)^2 + \frac{\sigma_5^2 + \sigma_6^2}{T^2} = 1 \qquad (6.5-14)$$

Compressive:

$$\sigma_1 = \frac{G_r}{1-k} \qquad (6.5-15)$$

where G_r is resin modulus and k is volume fraction ratio.

Matrix failure

Tensile:

$$\frac{(\sigma_2+\sigma_3)^2}{R^2}+\frac{(\sigma_4^2+\sigma_5^2+\sigma_6^2-\sigma_2\sigma_3)}{T^2}=1 \qquad (6.5-16)$$

Compressive:

$$\frac{1}{Y_C}\left[\left(\frac{Y_C}{2T}\right)^2-1\right](\sigma_2+\sigma_3)+\frac{(\sigma_2+\sigma_3)^2}{4T^2}+\frac{(\sigma_4^2+\sigma_5^2+\sigma_6^2-\sigma_2\sigma_3)}{T^2}=1 \qquad (6.5-17)$$

Delamination

$$\left(\frac{\sigma_3}{Z_T}\right)^2+\frac{\sigma_6^2+\sigma_5^2}{R^2}=1 \qquad (6.5-18)$$

A number of case studies involving first–ply failures and post–first–ply failures (or progressive failures) are presented in the book by Ochoa and Reddy [95]. For additional studies, the reader may consult the references listed below as well as the examples presented in the forthcoming chapters.

REFERENCES FOR ADDITIONAL READING

1. Garrett, K. W. and Bailey, J. E., "Multiple Transverse Fracture in 90° Cross–Ply Laminates of Glass Fibre–Reinforced Polyester," *Journal of Materials Science*, **12**, 157–168, 1977.

2. Highsmith, A. L. and Reifsnider, K. L., "Stiffness Reduction Mechanisms in Composite Laminates," *Damage in Composite Materials*, Reifsnider, K. L. (ed.), American Society for Testing and Materials, ASTM STP **775**, 103–117, 1982.

3. Talug, A., "Analysis of Stress Fields in Composite Laminates with Interior Cracks," Ph.D. Thesis, Virginia Polytechnic Institute and State University, Blacksburg, 1978.

4. Fukunaga, H., Chou, T. W., Peters, P. W. M., and Schulte, K., "Probabilistic Failure Strength Analyses of Graphite/Epoxy Cross–Ply Laminates," *Journal of Composite Materials,* **18**, 339-358, 1984.

5. Flaggs, D. L., "Prediction of Tensile Matrix Failure in Composite Laminates," *Journal of Composite Materials,* **19**, 29–50, 1985.

6. Laws, N. and Dvorak, G. J., "Progressive Transverse Cracking in Composite Laminates," *Journal of Composite Materials,* **22**, 900–916, 1988.

7. Hashin, Z., "Analysis of Cracked Laminates: A Variational Approach," *Mechanics of Materials,* **4**, 121–136, 1985.

8. Han, Y. M., Hahn, H. T., and Croman, R. B., "A Simplified Analysis of Transverse Ply Cracking in Cross–Ply Laminates," *Composites Science and Technology,* **31**, 165–177, 1988.

9. Lim, S. G. and Hong, C. S., "Prediction of Transverse Cracking and Stiffness Reduction in Cross–Ply Laminated Composites," *Journal of Composite Materials,* **23**, 695–713, 1989.

10. Kim, Y. M. and Hong, C. S., "Progressive Failure Model for the Analysis of Laminated Composites Based on Finite Element Approach," *Journal of Reinforced Plastics and Composites,* **11**, 1078–1091, 1992.

11. Zhang, J., Fan, J., and Soutis, C., "Analysis of Multiple Matrix Cracking in $[\pm\theta_m/90_n]_s$ Composite Laminates. Part 2 : Development of Transverse Ply Cracks," *Composites,* **23** (5), 299–304, 1992.

12. Lee, J. H. and Hong, C. S., "Refined Two–Dimensional Analysis of Cross–Ply Laminates with Transverse Cracks Based on the Assumed Crack Opening Displacement," *Composites Science and Technology,* **46**, 157–166, 1992.

13. Guild, F. J., Ogin, S. L., and Smith, P. A., "Modelling of 90° Ply Cracking Cross–Ply Laminates, Including Three–Dimensional Effects," *Journal of Composite Materials,* **27**(7), 646–667, 1993.

14. Nairn, J. A., "The Strain Energy Release Rate of Composite Microcracking: A Variational Approach," *Journal of Composite*

Materials, **23**, 1106–1129, 1989.

15. Herakovich, C. T., Aboudi, J., Lee, S. W., and Strauss, E. A., "Damage in Composite Laminates: Effects of Transverse Cracks," *Mechanics of Materials* **7**, 91–107, 1988.

16. Varna, J. and Berglund, L. A., "Multiple Transverse Cracking and Stiffness Reduction in Cross–Ply Laminates," *Journal of Composites Technology and Research,* **13**(2), 99–106, 1991.

17. Varna, J. and Berglund, L. A., "A Model for Prediction of the Transverse Cracking Strain in Cross–Ply Laminates," *Journal of Reinforced Plastics and Composites,* **11**, 708–728, 1992.

18. McCartney, L. N., "Theory of Stress Transfer in a $0° - 90° - 0°$ Cross–Ply Laminate Containing a Parallel Array of Transverse Cracks," *Journal of Mechanics and Physics of Solids,* **40**(1), 27–68, 1992.

19. Gamby, D. and Rebiére, J. L., "A Two–Dimensional Analysis of Multiple Matrix Cracking in a Laminated Composite Close to its Characteristic Damage State," *Composite Structures,* **25**, 325–337, 1993.

20. Talreja, R., "Stiffness Properties of Composite Laminates with Matrix Cracking and Interior Delamination," *Engineering Fracture Mechanics,* **25**(5,6), 751–762, 1986.

21. Allen, D. H., Harris, C. E., and Groves, S. E., "A Thermomechanical Constitutive Theory for Elastic Composites with Distributed Damage - I. Theoretical Development," *Int. J. Solids and Structures,* **23**(9), 1301–1318, 1987.

22. Allen, D. H. , Harris, C. E., and Groves, S. E., "A Thermomechanical Constitutive Theory for Elastic Composites with Distributed Damage - II. Numerical Results," *Int. J. Solids and Structures,* **23**(9), 1319–1338, 1987.

23. Lee, J. W., Allen, D. H., and Harris, C. E., "Internal State Variable Approach for Predicting Stiffness Reductions in Fibrous Laminated Composites with Matrix Cracks," *Journal of Composite Materials,* **23**, 1273–1291, 1989.

24. Tay, T. E. and Lim, E.H., "Analysis of Stiffness Loss in Cross–Ply Composite Laminates," *Composite Structures*, **25**, 419–425, 1993.

25. Wu, S. and Chudnovsky, A., "The Effective Elastic Properties of a Linear Elastic Solid with Microcracks," *Engineering Fracture Mechanics,* **37**(3), 653–659, 1990.

26. Gudmundson, P. and Östlund, S., "First–Order Analysis of Stiffness Reduction Due to Matrix Cracking," *Journal of Composite Materials,* **26**(7), 1009–1030, 1992.

27. Yuan, F. G. and Selek, M. C., "Transverse Cracking and Stiffness Reduction in Composite Laminates," *Journal of Reinforced Plastics and Composites,* **12**, 987–1015, 1993.

28. Kachanov, M., "Elastic Solids with Many Cracks and Related Problems," *Advances in Applied Mechanics,* J. W. Hutchinson and T. Y. Wu (eds.), **30**, 260–445, 1993.

29. Dvorak, G. J., Laws, N., and Hejazi, M., "Analysis of Progressive Matrix Cracking in Composite Laminates. I. Thermoelastic Properties of a Ply with Cracks," *Journal of Composite Materials,* **19**, 216–234, 1985.

30. Aboudi, J., "Stiffness Reduction of Cracked Solids," *Engineering Fracture Mechanics,* **26**(5), 637-650, 1987.

31. Sumarac, D. and Krajcinovic, D., "A Self–Consistent Model for Microcrack–Weakened Solids," *Mechanics of Materials* **6**, 39–52, 1987.

32. Bai, Jia–Min, "Stiffness Reduction Analysis of Cracked Cross–Ply Laminates by Using Integral Equation Approach," *Engineering Fracture Mechanics,* **34**(1), 245–251, 1989.

33. Aveston, J. and Kelly, A., "Theory of Multiple Fracture of Fibrous Composites," *Journal of Materials Science,* **8**, 352–362, 1973.

34. Brillaud, J. and El Mahi, A., "Numerical Simulation of the Influence of Stacking Sequence on Transverse Ply Cracking in Composite Laminates," *Composite Structures,* **17**, 23–35, 1991.

35. Caslini, M., Zanotti, C. and O'Brien, T. K., "Fracture Mechanics of Matrix Cracking and Delamination in Glass/Epoxy Laminates," NASA Technical Memorandum 89007, AVSCOM Technical Report 86-B-3, 1986.

36. Ogin, S. L., Smith, P. A., Beaumont, P. W. R., "Matrix

Cracking and Stiffness Reduction During the Fatigue of a $(0/90)_s$ GFRP Laminate," *Composite Science and Technology,* **22**, 23–31, 1985.

37. Wang, A. S. D., Chou, P. C., and Lei, S. C., "A Stochastic Model for the Growth of Matrix Cracks in Composite Laminates," *Journal of Composite Materials,* **18**, 239–254, 1984.

38. Praveen, G. N. and Reddy, J. N.,"Cross–Ply Laminates with a Periodic Array of Transverse Matrix Cracks. 1. Stress Transfer and Stiffness Reduction," Texas A&M University, College Station, 1995.

39. Praveen, G. N. and Reddy, J. N.,"Cross–Ply Laminates with a Periodic Array of Transverse Matrix Cracks. 2. Crack Opening Profiles and Stress Transfer at CDS," Texas A&M University, College Station, 1995.

40. O'Brien, T. K., "Characterization of Delamination Onset and Growth in a Composite Laminate," *Damage in Composite Materials*, K. L. Reifsnider (ed.), STP 775, American Society for Testing Materials, Philadelphia, 140–167, 1982.

41. O'Brien, T. K., "Analysis of Local Delaminations and Their Influence on Composite Laminate Behavior," *Delamination and Debonding of Materials*, STP 877, W. S. Johnson (ed.), American Society for Testing Materials, Philadelphia, 282–297, 1985.

42. O'Brien, T. K., "The Effect of Delamination on the Tensile Strength of Unnotched, Quasi–Isotropic, Graphite/Epoxy Laminates," *Proceedings of the SESA/JSME Joint Conference on Experimental Mechanics*, Part I, Honolulu, Hawaii, (SESA, Brookfield Center, CT), 236–243, 1982.

43. Rybicki, E. F., Schmueser, D. W., and Fox, J., "An Energy Release Rate Approach for Stable Crack Growth in the Free–Edge Delamination Problem," *Journal of Composite Materials,* **11**, 470–487, 1977.

44. Wang, A. S. D., and Crossman, F. W., "Initiation and Growth of Transverse Cracks and Edge Delamination in Composite Laminates, Parts 1 and 2," *Journal of Composite Materials*, Supplemental Volume, 71–106, 1980.

45. O'Brien, T.K., "Mixed-Mode Strain-Energy Release Rate Effects on Edge Delamination of Composites," *Effects of Defects in Composite Materials,* ASTM STP 836, American Society for Testing and Materials, 125–142, 1984.

46. Whitcomb, J. D. and Raju, I. S., "Analysis of Interlaminar Stresses in Thick Composite Laminates with and without Edge Delamination," *Delamination and Debonding of Materials*, ASTM STP 876, W. S. Johnson (ed.), American Society for Testing and Materials, Philadelphia, 69–94, 1985.

47. Chan, W. S. *et al.*, "Delamination Control of Composite Rotor Hubs," *Journal of the American Helicopter Society*, 60–69, 1986.

48. Raju, I. S., Crews, J. H., and Aminpour, M. A., "Convergence of Strain Energy Release Rate Components for Edge-Delaminated Composite Laminates," NASA Technical Memorandum 86135, 1987.

49. O'Brien, T. K. *et al.*, "Comparisons of Various Configurations of the Edge Delamination Test for Interlaminar Fracture Toughness," *Toughened Composites*, ASTM STP 937, N. J. Johnston (ed.), American Society for Testing and Materials, Philadelphia, 199–221, 1987.

50. Salpekar, S. A., and O'Brien, T. K., "Combined Effect of Matrix Cracking and Stress-Free Edge on Delamination," NASA Technical Memorandum 102591, March 1990.

51. Robbins, D. H. and Reddy, J. N., "The Effects of Kinematic Assumptions on Computed Strain Energy Release Rates for Delaminated Composite Plates," *Mathematical Modelling and Scientific Computing*, **1** (1,2), 50–66, 1993.

52. Barbero, E. J. and Reddy, J. N., "Modeling of Delamination in Composite Laminates Using a Layer-Wise Plate Theory," *Int. J. Solids and Structures*, **28**(3), 373–388, 1991.

53. Robbins, D. H. and Reddy, J. N., "Modeling of Thick Composites Using a Layerwise Laminate Theory," *Int. J. Numer. Meth. Engng.,* **36**, 655–677, 1993.

54. Whitney, J. M., Browning, C. E., and Hoogsteden, W., "A Double Cantilever Beam Test for Characterizing Mode I Delamination of Composite Materials," *Journal of Reinforced*

Plastics and Composites, **1**, 297–313, 1982.

55. Crews, J. H., Jr., Shivakumar, K. N., and Raju, I. S., "Factors Influencing Elastic Stresses in Double Cantilever Beam Specimens," NASA Technical Memorandum 89033, 1987.

56. Martin, R. M., "Effect of Initial Delamination on GIc and GI Values from Glass/Epoxy Double Cantilever Beam Tests," *Proceedings of the American Society for Composites*, Third Technical Conference, Seattle, Washington, Sept. 25–29, 1988,

57. Raju, I. S., Shivakumar, K. N., and Crews, J. H., "Three–Dimensional Elastic Analysis of a Composite Double Cantilever Beam Specimen," *AIAA Journal*, **26** (12), 1493–1498, 1988.

58. Davidson, B. D., and Schapery, R. A., "Effect of Finite Width on Deflection and Energy Release Rate of an Orthotropic Double Cantilever Beam Specimen," *Journal of Composite Materials*,**22**, 640–656, 1988.

59. Crews, J. H., Jr., Shivakumar, K. N., and Raju, I. S., "Effects of Anticlastic Curvature on *G* Distributions for DCB Specimens," *AIAA/ASME/ASME/ASCE/AHS/ASC Structures, Structural Dynamics and Materials Conference, 30th*, Mobile, Alabama, April 3–5, 1989.

60. Krishna Kumar, R. and Reddy, J. N., "Stress Distributions During Fiber Pull Out," *Journal of Applied Mechanics,* in press.

61. Kim, J. K. and Mai, Y. W., "High Strength, High Fracture Toughness Fiber Composites with Interface Control - A Review," *Composite Sci. Tech.*, **41**, 333–378, 1991.

62. Suzuki, T. *et al.*, "Fiber Pullout Processes and Mechanisms of Carbon Fiber Reinforced Silicon Nitride Ceramic Composite," *J. Mat. Res.*, **7**(10), 2809–2875, 1992.

63. Budiansky, B. *et al.*, "Matrix Fracture in Fiber–Reinforced Ceramics," *J. Mech. Phys. Solids*, Vol. 34, 1986, 167–189.

64. Desarmot, G. and Favre, J. P., "Advances in Pull–Out Testing and Data Analysis," *Compos. Sci. Tech.*, **42**, 151–187, 1991.

65. Khanna, K. S., "Effect of Reinforcement and Fiber–Matrix Interface on Dynamic Fracturing of Fiber Reinforced

Composite Materials," Ph.D. Thesis, University of Rhode Island, Kingston, RI, 1992.

66. Atkinson, C. *et al.*, "The Rod Pull Out Problem, Theory and Experiment," *J. Mech. Phys. Solids*, **30**(3), 97–120, 1982.

67. Bowling, J. and Groves, G. W., "The Debonding and Pull–Out of Ductile Wires from a Brittle Matrix," *J. Mat. Sci.*, **14**, 431–442, 1979.

68. Favre, J. P. and Menenne, M. C., "Characterization of Fiber/Resin Bonding in Composites Using a Pull–Out Test," *Int. J. Adhesion and Adhesives,* **1**, 311–316, 1981.

69. Hutchinson, J. W. and Jensen, H. M., "Models of Fiber Debonding and Pull Out in Brittle Composites with Friction," *Mech. Mater.*, **9**, 139–163, 1990.

70. Aboudi, J., "Micromechanical Analysis of Fibrous Composites with Coulomb Frictional Slippage Between the Phases," *Mech. Mater.*, **8**, 103–115, 1989.

71. Gao, Y. C. *et al.*, "Fracture of Fiber Reinforced Materials," *J. Appl. Math and Phys. (ZAMP),* **39**, 550–572, 1988.

72. Dollar, A. and Steif, P. S., "Load Transfer in Composites with a Coulomb Friction Interface," *Int. J. Solids Struct.*, **24**(8), 789–803, 1988.

73. Fruend, L. B., "The Axial Force Needed to Slide a Circular Fiber Along a Hole in an Elastic Material and Implications for Fiber Pull–Out," *Eur. J. Mech. Solids,* **11**, 1–19, 1992.

74. Povirk, G. L., and Needleman, A., "Finite Element Simulations of Fiber Pull–Out," *Trans. ASME, J. Eng. Mater. Tech.*, **115**, 286–291, 1993.

75. Ballarami, R. and Ahmed, S.,"Finite Element Modeling of Frictionally Restrained Composite Interfaces," NASA Contract report 182281, 1989.

76. Marotzke, "The Elastic Stress Field Arising in the Single Fiber Pull–Out Test," *Comp. Sci. Tech.*, **50**, 393–403, 1994.

77. Faber, K. T. *et al.*, "Frictional Stress Evaluation Along the Fiber–Matrix Interface in CMC's," *J. Amer. Ceram. Soc.*,

69(9), 208–209, 1986.

78. Hsueh, C. H., "Interfacial Debonding and Fiber Pull–Out Stresses of Fiber Reinforced Composites, X: with an Elastic Interfacial Coating", *Mat. Sci. Engng.*, **105**, 189–195, 1993.

79. Reddy, J. N. and Pandey, A. K., "A First-Ply Failure Analysis of Composite Laminates," *Computers and Structures*, **25** (3), 371–93, 1987.

80. Turvey, G. J. and Osman, M. Y., "Exact and Approximate Linear and Nonlinear Initial Failure Analysis of Laminate Mindlin Plates in Flexure," *Composite Structures* **5**, I. H. Marshall (ed.), Elsevier Applied Science, London, Chapter 4, 133–171, 1989.

81. Reddy, Y. S. N. and Reddy, J. N., "Linear and Non-Linear Failure Analysis of Composite Laminates with Transverse Shear," *Composites Science and Technology*, **44**, pp. 227–255 (1992).

82. Tsai, S. W., "A Survey of Macroscopic Failure Criteria for Composite Materials," Technical Report, AFWAL-TR-84-4025.

83. Soni, S. R., "A Comparative Study of Failure Envelopes in Composite Laminates," *Journal of Reinforced Plastics and Composites*, **2** (1), 34–42, 1983.

84. Hashin, Z., "Failure Criteria for Unidirectional Composites," *Journal of Applied Mechanics*, **47**, pp. 329-334 (1980).

85. Petit, P. H. and Waddoups, M. E., "A Method of Predicting the Nonlinear Behavior of Laminated Composites," *Journal of Composite Materials*, **3** (Feb. 1969).

86. Sandhu, R. S., Sendeckyj, G. P., and Gallo, R. L., "Modeling of the Failure Process in Notched Laminates," in *Mechanics of Composite Materials Recent Advances,* Proceedings of the IUTAM Symposium held at Virginia Polytechnic Institute and State University, August 16–19, 1982, Z. Hashin and C. T. Herakovich (eds.), Pergamon Press, Elmsford, NY, 179–189, 1982.

87. Tsai, S. W. and Azzi, V. D., "Strength of Laminated Composite Materials," *AIAA Journal*, **4** (2), pp. 296–301 (Feb. 1966).

88. Chang, F. K. and Chang, K. Y., "A Progressive Damage Model for Laminated Composites Containing Stress Concentrations," *Journal of Composite Materials,* **21**, 1987.

89. Chang, F. K., Lessard, L., and Tang, J. M., "Compression Response of Laminated Composites Containing an Open Hole," *SAMPE Quarterly* **19**, 1988.

90. Hahn, H. T. and Tsai, S. W., "Nonlinear Elastic Behavior of Unidirectional Composite Laminates," *Journal of Composite Materials,* **7**, 1973.

91. Yamada, S. E. and Sun, C. T., "Failure Strength of Nonlinearly Elastic Composite Laminates Containing a Pin Loaded Hole," *Journal of Composite Materials,* **18**, 1984.

92. Tan, S. C., "A Progressive Failure Model of Composite Laminates Containing Openings," *Journal of Composite Materials,* **25**, 556-577, 1991.

93. Starnes, J. H., Jr. and Rouse, M., "Postbuckling and Failure Characteristics of Selected Flat Rectangular Graphite-Epoxy Plates Loaded in Compression," AIAA Paper No. 81-0543 (1981).

94. Engelstad, S. P., Reddy, J. N., and Knight, N. F., Jr., "Postbuckling Response and Failure Prediction of Graphite–Epoxy Plates Loaded in Compression," *AIAA Journal*, **30**(8), 2106–2113, 1992.

95. Ochoa, O. O. and Reddy, J. N., *Finite Element Analysis of Composite Laminates*, Kluwer Academic Publishers, The Netherlands, 1992.

96. Nuismer, R. J. and Tan, S. C., "Constitutive Relations of a Cracked Composite Lamina," *J. Comp. Mater.*, **22**, 306–321, 1988.

Chapter Seven

VARIABLE THICKNESS STRUCTURES

7.1 INTRODUCTION

Variable thickness structures are a natural choice because of the saving in material and hence cost. Since most composite laminates are composed of a number of layers, it is possible to construct variable thickness composite structures by terminating each ply in such a way as to meet the design requirements (see Figure 7.1–1).

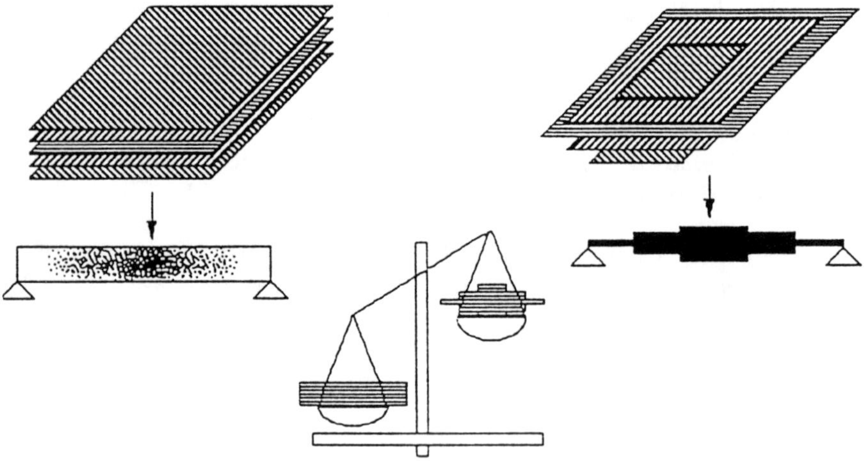

Figure 7.1–1: Schematic representation of a variable thickness composite structure and comparison with a constant thickness one.

A general study of variable thickness beams and plates subjected to transverse loads is presented in this chapter. First of all, the theoretical background for obtaining general variable thickness structures is described. The stress concentration problem is also analyzed. Special attention is paid to stress distributions through the laminate thickness. A sensitivity study is carried out in order to determine the critical parameters in terms of stress concentrations. Some examples of applications to beams and plates are presented for different types of boundary conditions. Finally, a theoretical–experimental correlation is shown to assess the accuracy of the practical analysis methodologies described in the present chapter.

7.2 THEORETICAL BACKGROUND

The recurrence relations used here for resizing the elements are based on optimality criteria, which are not rigorous in the mathematical sense but which are found to give near optimum weight designs for large structures in a very efficient way.

The optimality criterion for the generalized stiffness requirement can be stated as, "the optimum structure is the one in which the ratio of the average strain energy density to the mass density is the same in all the elements." This criterion can be written as:

$$1 = \lambda\,\frac{\tilde{e}_{ij}}{\rho_{ij}} \quad \begin{cases} i = 1, \cdots, m \\ j = 1, \cdots, m \end{cases} \tag{7.2 - 1}$$

where λ is the Lagrangian parameter, $\tilde{e}_{ij}$ is the strain energy density of the jth layer of the ith element, and ρ_{ij} is the mass density. The strain energy density is given by:

$$\tilde{e}_{ij} = \frac{e_{ij}}{V_{ij}} \tag{7.2 - 2}$$

where e_{ij} is the strain energy, and V_{ij} is the volume of the jth layer of the ith element. The volume of the element defined by

$$V_{ij} = t_{ij}\; I_j \tag{7.2 - 3}$$

where t_{ij} is the thickness of the jth layer of the ith element, and I_j is the surface area of the ith element. The design variable t_{ij} can be written as

$$t_{ij} = \Lambda \alpha_{ij} \tag{7.2 – 4}$$

where α_{ij} is the relative thickness (normalized to the maximum thickness) of the jth layer of the ith element, and Λ is the scaling parameter

$$\Lambda[K']\{a\} = \{q\} \tag{7.2 – 5}$$

or

$$[K']\{a'\} = \{q\} \tag{7.2 – 6}$$

where

$$\{a'\} = \Lambda\{a\} \tag{7.2 – 7}$$

and $[K']$ is the stiffness matrix for the whole structure obtained by using the relative design vector α_{ij}. Introducing the scaling parameter into the finite element equations, the relations between the actual quantities and the relative quantities at element level can be expressed as

$$[k]_{ij} = \Lambda[k']_{ij} \tag{7.2 – 8}$$

$$\{a\}_i = \frac{1}{\Lambda}\{a'\}_i \tag{7.2 – 9}$$

$$\{\varepsilon\}_i = \frac{1}{\Lambda}\{\varepsilon'\} \tag{7.2 – 10}$$

$$\{\sigma\}_{ij} = \frac{1}{\Lambda}\{\sigma'\}_{ij} \tag{7.2 – 11}$$

where the quantities with a prime denote the relative values.

Introducing Equations (2), (8) and (9) in Eq. (1) gives

$$1 = \frac{\lambda}{\Lambda^2} \frac{\tilde{e}'_{ij}}{\rho_{ij}} \qquad (7.2-12)$$

where

$$\tilde{e}'_{ij} = \frac{1}{\alpha_{ij} 1_i} \{r'\}_i^T [K']_{ij} \{r'\}_i \qquad (7.2-13)$$

Multiplying both sides of equations (12) by α_{ij}^2 and taking the square root gives

$$\alpha_{ij} = B\alpha_{ij} \left(\tilde{e}'_{ij} / \rho_{ij} \right)^{\frac{1}{2}} \qquad (7.2-14)$$

where B is the constant, Equation (14) can be rewritten in an iterative form as

$$(\alpha_{ij})_{\nu+1} = B(\alpha_{ij})_{\nu} \left(\tilde{e}'_{ij} / \rho_{ij} \right)_{\nu}^{\frac{1}{2}} \qquad (7.2-15)$$

where ν refers to the iteration number.

In this procedure the resizing of an element is done by dividing the design variable by the minimum actual strength/stress ratio for that element:

$$(\alpha_{ij})_{\nu+1} = (\alpha_{ij})_{\nu} / (r_{min})_{\nu} \qquad (7.2-16)$$

Therefore, the selection of the appropriate failure criterion (see Chapter 6) is critical. If the strength/stress ratio defined by the failure criterion is unique for each combined state of stress and the corresponding state of strain, the application of Eq. (16) is efficient and simple. Otherwise, the application of the procedure becomes more complicated. Thus, a quadratic failure criterion discussed in Section 6.5 is recommended.

This iterative procedure can be implemented into a general finite element program.

7.3 THE EFFECT OF STRESS CONCENTRATION

7.3–1 Introduction

The aim of this study is to find out what happens inside the laminate when the laminate thickness varies and a transverse load is applied. The behavior of a variable thickness plate structure is extremely difficult to understand because, in general, it is a 3–D problem, and bending, shear, and variable thickness effects appear simultaneously. In other words, all six stress components must be considered. The three in–plane stress components are present owing to the bending effects. The two interlaminar shear stress components present because of the shear effects. Finally, the interlaminar normal stress component must also be considered because of the variable thickness effect. In order to simplify the problem and to understand the variable thickness phenomenon, a one–dimensional variable thickness plate (*i.e.*, a plate strip or beam) is studied using a 2–D plane strain finite element model. The domain modeled is the thickness of the plate strip.

In the next section, we discuss the attempts made in obtaining through–thickness stresses distributions in a variable cross section (*i.e.*, the laminate thickness varies). Once the stress level is known, the laminate failure is predicted by means of a quadratic failure criterion (see Chapter 6). The aim of this study is to analyze theoretically the influence of different parameters on the mechanical behavior of variable thickness laminated composite plates. An investigation is carried out to study the damage in variable thickness laminated composite plates subjected to a transverse load. The main objective of the study is to gain a fundamental understanding of the failure mechanisms caused by transverse loads and to identify the essential parameters causing the damage in composites. Finally, an experimental study is carried out in order to assess the accuracy of the theory, as well as to analyze experimentally the influence of different parameters on the mechanical behavior of variable thickness laminated composite plates.

7.3–2 Stress Distributions

The model used here is based on the following assumptions:

- This study is limited to one–dimensional laminated composite plates in order to avoid three–dimensional effects and aims to analyze properly the consequences of the variable thickness effect.

- The fiber orientation is longtitudinal (x–direction). This is the optimal direction for one–dimensional laminated composite plates.

- The material used is graphite/epoxy AS4/3501–6.

The method of analysis is characterized by the following features:

- A 2–D plane strain finite element model is used.

- Four–node, linear, quadrilateral finite elements are used.

- The mesh is composed of 1891 nodes.

- Different types of loading and boundary conditions are studied to analyze the variable thickness problem from a theoretical point of view and to obtain through–thickness stress distributions.

- A three–point bending model is used for analyzing theoretically the influence of the different parameters on the mechanical behavior of variable thickness laminated composite plates.

- A three–point bending test in fatigue and static conditions is carried out for identifying the failure mechanisms.

- A three–point bending test in static conditions is used to verify the theoretical model and to analyze experimentally the influence of the different parameters on the mechanical behavior of variable thickness laminated composite plates.

In Figures 7.3–1 to 7.3–3, σ_1, σ_3 and σ_5 distributions are represented for $t_1/t = 0.3$, $M/Pt = 5$ and $\partial M/\partial x < 0$, respectively, where M denotes the bending moment, P the transverse force, and t is the total thickness. Two angles of variation of thickness have been considered: 15° and 90°. The reduction in the angle of variation of thickness from 90° to 15° leads to a considerable decrease in the peak values of the distributions of σ_1, σ_3 and σ_5.

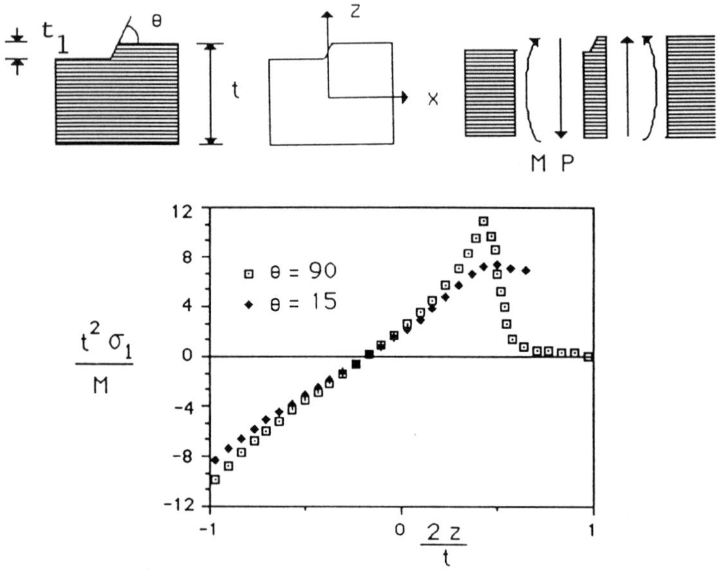

Figure 7.3–1: Distribution of σ_1 for a one–side tapered plate ($t_1/t = 0.3, M/Pt = 5$ and $\frac{\partial M}{\partial x} > 0$).

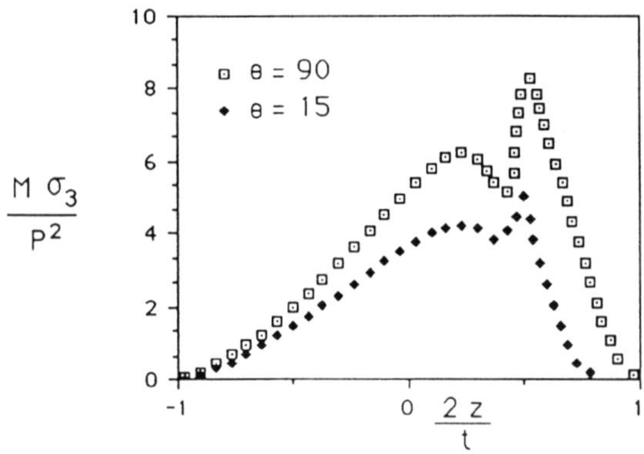

Figure 7.3–2: Distribution of σ_3 for a one–side tapered plate ($t_1/t = 0.3, M/Pt = 5$ and $\frac{\partial M}{\partial x} > 0$).

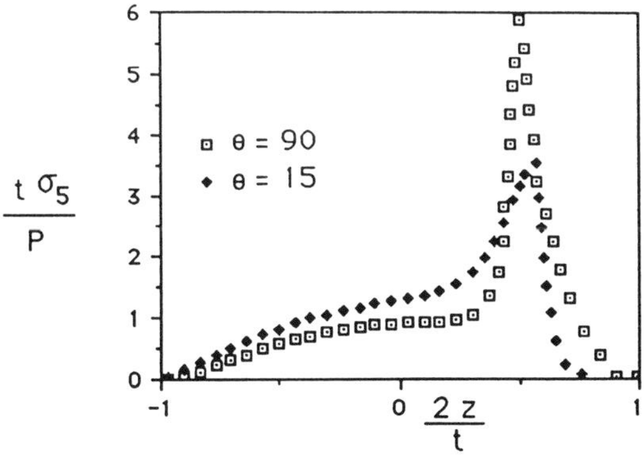

Figure 7.3–3: Distribution of σ_5 for a one–side tapered plate ($t_1/t = 0.3, M/Pt = 5$ and $\frac{\partial M}{\partial x} > 0$).

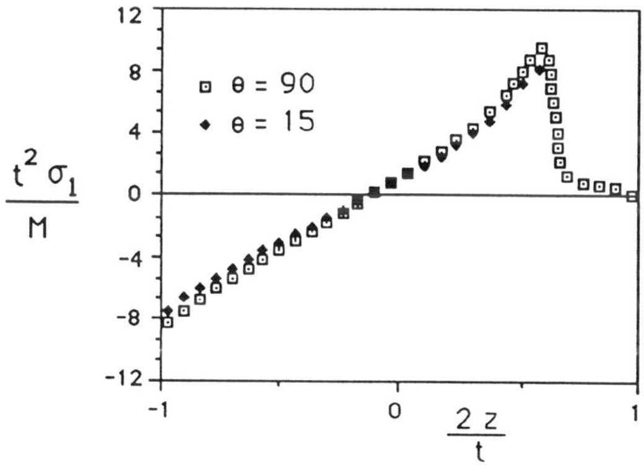

Figure 7.3–4: Distribution of σ_1 for a one–side tapered plate ($t_1/t = 0.2, M/Pt = 5$ and $\frac{\partial M}{\partial x} > 0$).

Figures 7.3–4 through 7.3–6 show the distributions of σ_1, σ_3, and σ_5, respectively, for $t_1/t = 0.2, M/Pt = 5$ and $\partial M/\partial x < 0$. As expected, the peak values decrease considerably with respect to the results in Figures 7.3–1 through 7.3–3 because of the smaller value of t_1/t.

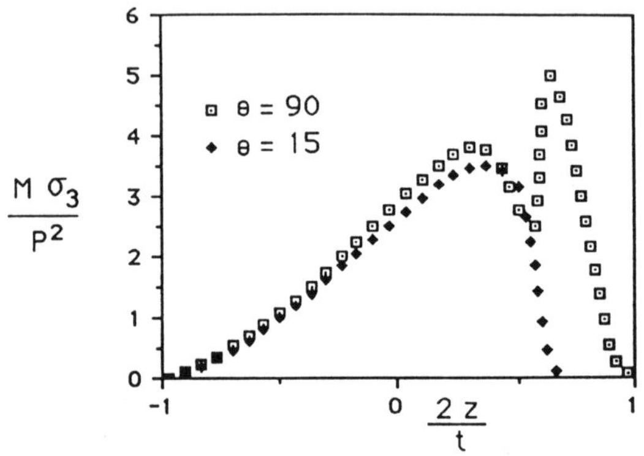

Figure 7.3–5: Distribution of σ_3 for a one–side tapered plate ($t_1/t = 0.2, M/Pt = 5$ and $\frac{\partial M}{\partial x} > 0$).

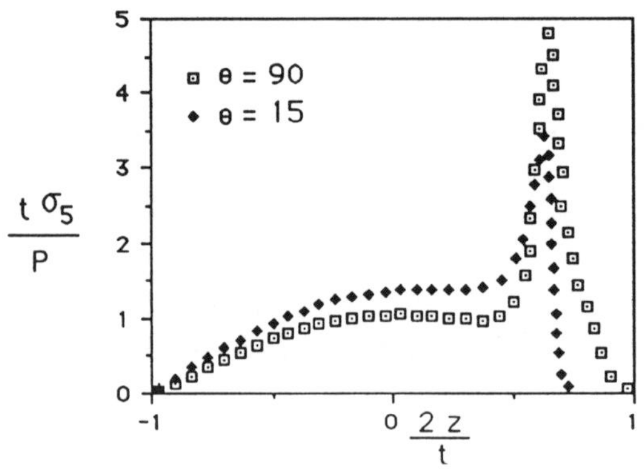

Figure 7.3–6: Distribution of σ_5 for a one–side tapered plate ($t_1/t = 0.2, M/Pt = 5$ and $\frac{\partial M}{\partial x} > 0$).

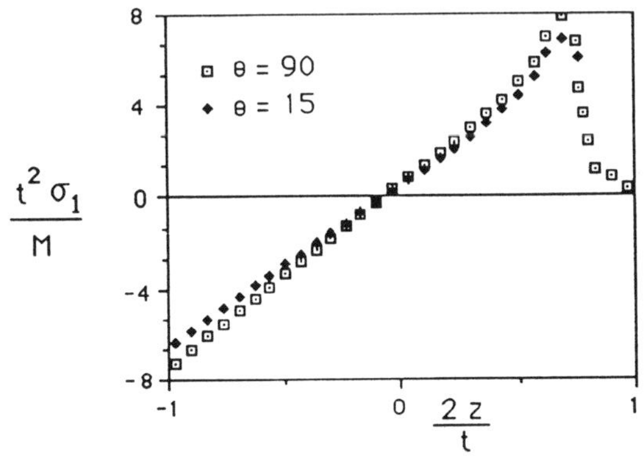

Figure 7.3–7: Distribution of σ_1 for a one–side tapered plate $(t_1/t = 0.1, M/Pt = 5$ and $\frac{\partial M}{\partial x} > 0)$.

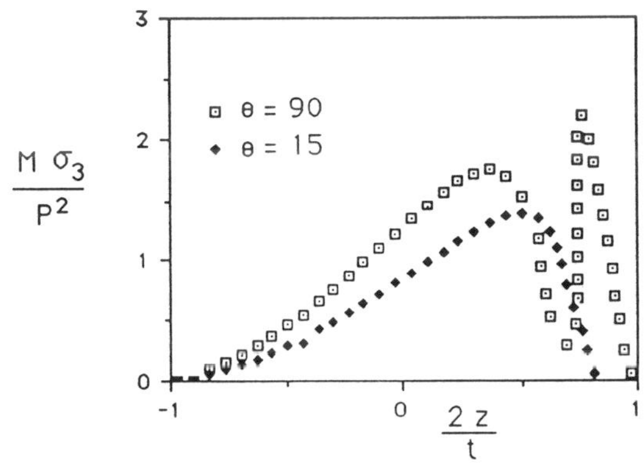

Figure 7.3–8: Distribution of σ_3 for a one–side tapered plate $(t_1/t = 0.1, M/Pt = 5$ and $\frac{\partial M}{\partial x} > 0)$.

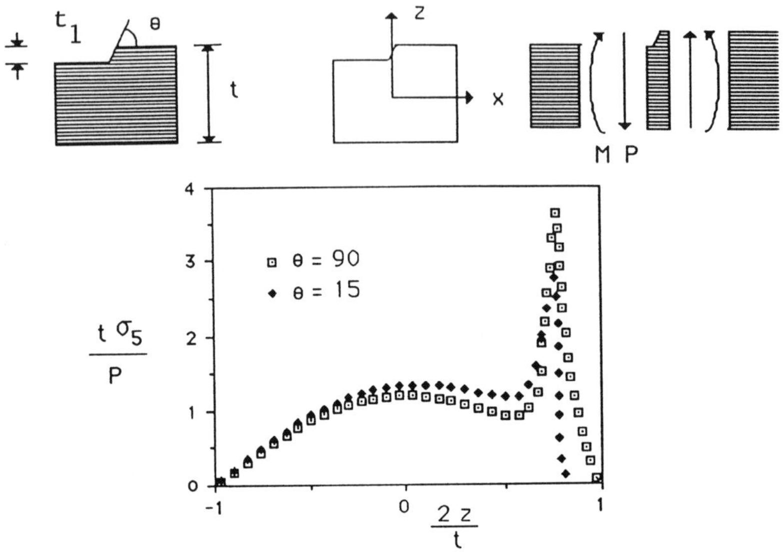

Figure 7.3–9: Distribution of σ_5 for a one–side tapered plate ($t_1/t = 0.1, M/Pt = 5$ and $\frac{\partial M}{\partial x} > 0$).

Distributions of σ_1, σ_3, and σ_5 for $t_1/t = 0.1, M/Pt = 5$ and $\partial M/\partial x < 0$ are presented in Figures 7.3–7 through 7.3–9, respectively. The peak values keep decreasing because of the value of $t_1/t = 0.1$. Also, the effect of the angle of variation of thickness can be seen by comparing the distributions for $\theta = 90°$ and $\theta = 15°$.

The following conclusions can be drawn after analyzing the results presented in the figures above:

- Stress distributions are strongly dependent on the value of the thickness ratio t/t_1, as indicated by the results in Figures 7.3–1 through 7.3–9.

- The angle of variation of thickness, θ, has a remarkable influence on interlaminar stress distributions through the laminate thickness. This fact can be seen in Figures 7.3–1 through 7.3–9. In general, values of $\theta = 90°$ yield the larger peak values of the stress compared to $\theta = 15°$.

7.3–3 Failure Mechanisms

A variable thickness laminated composite plate exhibits a variety of failure mechanisms when subjected to transverse loads. The knowledge of these modes is very important for the understanding of what happens inside the laminate. Once we know the critical failure mechanism, we will be able to identify the strain components associated with them.

Different types of laminated composite plates were tested in fatigue and static conditions:

- AS4/3501–6 graphite/epoxy was used for this work.
- The types of testing performed were three–point static and dynamic bending.
- The load rate for the static testing was 0.06 in/min (0.001524 m/min).
- For the fatigue testing, the frequency was 5 cycles/second, and the load was 60% of the static failure load. The transverse displacement was controlled.

After studying the results, the following failure mechanisms have been identified:

1. *Delamination.* This is the most usual failure mode when the angle of variation of thickness is high. It is due to the fact that a sharp thickness variation generates a peak value of interlaminar shear stress near the areas of change of thickness. This mechanism can also be detected in thick plates, though the angle of variation of thickness is low.

2. *Compression Bending Modes.* This mechanism occurs when the plate is very thin, the variable thickness effect is negligible, and the compression strength is lower than the tension strength in the fiber direction. The following in–plane compression failure modes can be distinguished.

   2.1 Transverse tensile failure. Unidirectional composites can fracture along the fibers when loaded by compression by a transverse tensile failure mode, because of the weakness of the matrix and the fiber–matrix interface, compared with the strength of the fibers.

2.2 Compressive delamination failure. If a fiber buckles, the fiber–matrix interface may fracture in shear and lead to ultimate failure.

2.3 Euler failure. If the matrix is ductile and the interface is strong, the fiber can bend without matrix failure and eventually fracture in bending.

2.4 Microbuckling. A more likely failure mode of unidirectional composite laminates associated with fiber microbuckling and fiber kinking, is shear crippling. Macroscopically, shear crippling looks like a shear failure on a plane at an angle to the direction of loading. Microscopic inspection, however, indicates that shear crippling is frequently the result of kink–band formation.

2.5 Strength failure. The final failure mode exhibited in unidirectional composites is associated with pure compression failure of the fibers. In this case, the fracture surface is likely to be at an angle to the loading direction, usually about 45°.

3. *Tension Bending Modes.* This mechanism occurs when the plate is very thin, the variable thickness effect is negligible, and the compression strength is higher than the tension strength in the fiber direction. The following in–plane tension failure modes can be distinguished:

3.1 Brittle failure. Unidirectional composite subjected to longitudinal tensile load. In this mode, stress concentrations created at the broken fiber ends will lead to specimen separation at a given cross section.

3.2 Brittle failure with fiber pull–out. Unidirectional composite subjected to longitudinal tensile load. Variations in bond strength and local load transfer mechanisms from matrix to fiber can lead to the pull–out of the fibers from the matrix at fracture.

3.3 Brittle failure with debonding and/or matrix failure. Unidirectional composite subjected to longitudinal tensile load. Finally, in other cases, cracks at different cross sections of the laminate may join together at fracture through fiber–matrix debonding or by shear failure of the matrix. This inter–fiber–matrix shear failure and fiber–matrix debonding can occur either independently or in combination; that is, portions of the failure path may exhibit debonding, while matrix shear failure is evident in other regions.

4. *Mixed Modes.* A mixed failure mechanism has been detected in the regions close to the applications of the load. There exists a compressive delamination failure due to the bending mode (compression) and crushing due to a compressive peeling or interlaminar normal stress. This mechanism was observed in the three–point bending fatigue test, after 50,000 cycles. At the top, a crushing failure due to compressive interlaminar normal stress is shown. This mixed mode does not appear in the four–point bending test, and it can also be avoided by placing a hard rubber pad (60–80 durometer) between the load point and the specimen.

After analyzing the failure mechanisms described above, the following conclusions can be drawn:

- The critical failure modes are related to two strain components: the longitudinal normal strain ε_1 and interlaminar shear strain ε_5.

- The bending (compression or tension) modes are likely to occur when the plate is thin, and the variable thickness effect is negligible.

- The delamination mechanisms are usual in cases where the variable thickness effect is critical (sharp change of thickness). The failure appears in the tapered surface. This mechanism also occurs when the plate is thick, even if the variable thickness is negligible. In this case, the failure appears near the middle plane of the laminate.

- The mode related to interlaminar normal strains (mixed modes) is due to stress concentrations in the areas near the application of the load. This mechanism can be easily avoided by means of the solutions mentioned above.

7.4 EXAMPLES

7.4–1 Preliminary Comments

In this section, a number of numerical results for one– and two–dimensional laminated composite plates are described. First, the method of analysis used is outlined. The conclusions drawn in the last section are the key for the formulation of the present problem. Owing to the fact that the results obtained from the plane strain model are very close to the data given by the experimental study, this model can be directly applied for analyzing one–dimensional laminated composite plates. The model can also be the basis of a more general one, applicable to the analysis of two–dimensional laminated composite plates.

Once the method of analysis is known, the input data required for the calculation must be obtained. There are two types of input data in a structural optimization problem: geometric and material. There is no problem in getting the geometric data of the plate structure, but the elastic constants and strengths of the material require a special effort. In–plane strengths can be found in the literature, although some doubts exist about the compression strength. However, through–thickness properties are hard to get. Interlaminar normal and interlaminar shear moduli are a function of the stacking sequence. Therefore, the elastic constants and strengths are obtained experimentally for the materials used.

7.4–2 Model Assumptions and Method of Analysis

There are two alternatives to optimize plate structures:

- Use a 3–D theory and a three–dimensional finite element model.
- Use a 2–D plate theory and a shear deformation plate finite element model.

The first option is expensive, especially from the point of view of the optimization and subsequent application of an iterative procedure. The second option is efficient, because a 2–D plate finite element is computationally less expensive than a full 3–D finite element. Analysis can be carried out very fast because the total number of degrees of freedom is not large. However, the following question remains to be answered: is a 2–D shear deformation plate theory accurate enough to study variable thickness plate structures? This question has two possible answers:

- If the angle of variation of thickness is large (close to 90°), there are discontinuities in the exterior surface(s) of the laminated composite plates, there are free edge effects, and a 3–D finite element theory is required to resolve these effects.

- If the angle of variation of thickness is small, and the plate exterior surfaces are continuous, a bidirectional plate with variable thickness can be studied very accurately by means of a shear deformation laminated plate theory (see Chapters 4 and 5). This fact is based on a number of verifications made between plane strain models and models using a shear deformation plate theory. In all cases, deflections were very close and, according to the conclusions of failure mechanisms, the critical stress components σ_1 and σ_5 showed very small differences.

Both theoretical and experimental studies of variable thickness plates showed that the angle of variation of thickness has a strong influence on the failure load, and that the lower the angle the higher the failure load. Hence, it is logical to design tapered plates with low angles of variation of thickness and continuous exterior surfaces. In this case, a shear deformation plate theory is applicable.

The finite element used here is based on the third–order shear deformation theory of Reddy [1] and the penalty function method. This model makes it possible to analyze thin and thick plates, due to its general formulation. The results obtained by the finite element method are optimized by means of an iterative procedure. The laminate thickness and fiber orientation are the design variables. Laminate thickness is modified in each step by means of an iterative procedure based on optimality criteria to design a minimum weight structure. In order to assess the stress level, a quadratic failure criterion is applied in each element. If the stresses in all the elements satisfy this criterion, the process is over. If they do not, another iteration starts. Usually, around ten iterations are needed to get convergence.

Optimization of fiber orientations is complex due to the following reasons:

- Owing to practical considerations, only four angles have been used: $0°, 45°, -45°$ and $90°$.
- Owing also to practical considerations, the optimum laminate should be defined as a function of repetitive sublaminates.
- The stacking sequence has a strong influence on the strain and stress level, due to bending and interlaminar effects.

No acceptable results were obtained by using the optimization procedures available. Finally, the following assumptions were made in the development of the analysis:

- The analysis is static.
- The theory applied is valid for small deformations.
- The material is elastic.

7.4–3 Laminated Composite Beams

The problem was analyzed using the following data:

- The objective is to obtain the minimum weight structure by using strength as a design criterion.
- The laminate used is $[0]_n$ because it is the optimum for one–dimensional laminated composite plates subjected to transverse loads.
- The laminate is tapered on both sides of the midplane.
- Three kinds of plates have been analyzed:
  - thin plates ($a/t >> 10$).
  - thick plates ($a/t << 10$).

  - sandwich panels.

- Boundary conditions are restricted to:
  - simply supported.
  - clamped.
  - cantilevered.

- Two types of loading are applied:
  - uniform load.
  - point load at the center of the laminate for simply supported and clamped laminate, and at the free end of the laminate for cantilevered laminate.

- The mesh was composed of 400 nodes.

- Four–node, linear, quadrilateral finite elements were used.

- The material used is T300/N5208.

The optimal thickness thin and thick cantilevered plates under uniformly distributed transverse load are discussed here.

Thin beams subjected to a uniform load

Optimal thin plates subjected to a uniform load are represented in Figure 7.4–1. The optimum laminate thickness in all cases is given by

$$t = 2ka\sqrt{\frac{P}{X}} \qquad (7.4-1)$$

where a is the laminate length, P is the load applied, X is the uniaxial tensile strength of a ply along the x–axis, and k is a coefficient that varies as a function of x. The variation of the non–dimensional coefficient k is shown in Figure 7.4–1 for the three boundary conditions. Also, the weight savings (WS) are reported for these three cases. The 45.7% weight saving for the cantilever plate under uniformly distributed transverse load is remarkable.

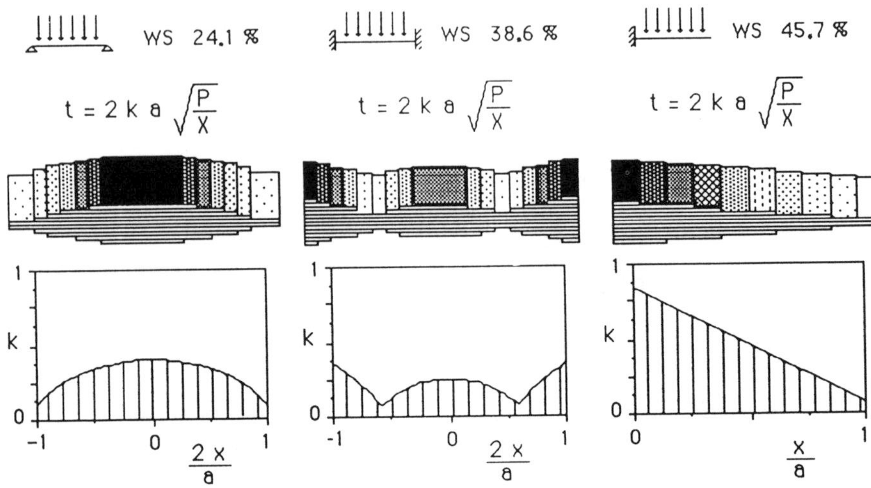

Figure 7.4–1: Representation of optimal thin one–dimensional laminates (*i.e.*, beams) subjected to a uniform distributed transverse load.

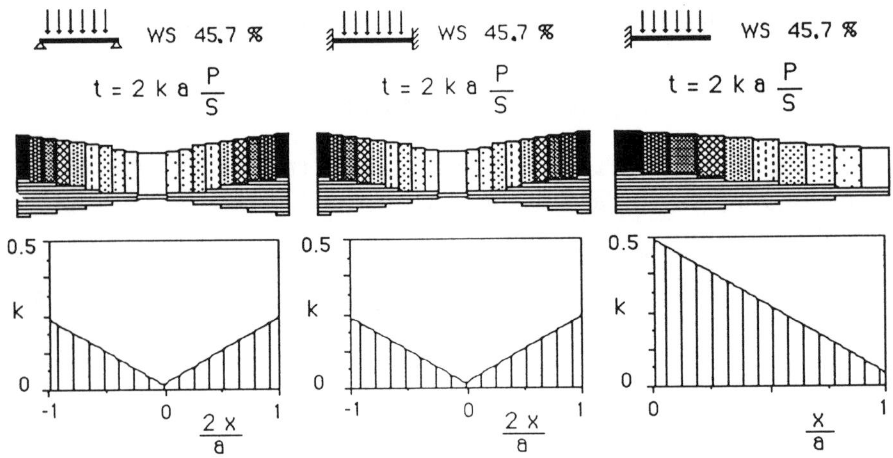

Figure 7.4–2: Representation of optimal thick one–dimensional laminates (*i.e.*, beams) subjected to a uniform distributed transverse load.

Thick beams subjected to a uniform load

Figure 7.4–2 shows the optimal thick plates subjected to a uniform load. In this case, the optimum laminate thickness is

$$t = 2ka\frac{P}{S} \qquad (7.4-2)$$

where S is the shear strength in the $xy-$ or 12–plane of a ply. The value of the nondimensional coefficient k as a function of x and the optimal beam configurations are given for the three boundary conditions mentioned above. The weight saving is the same for all three boundary conditions, namely 45.7%.

7.4–4 Laminated Composite Plates

The problem analyzed here is formulated using the following points:

- The objective is to get the minimum weight structure by using strength as a design criterion.
- Only symmetric and balanced laminates are considered.
- Symmetric taper about the midplane is considered.
- Three kinds of plates are analyzed:
  - thin plates $(b/t >> 10)$.
  - thick plates $b/t << 10)$.
  - sandwich panels.
- Boundary conditions are restricted to:
  - simply supported on all four edges.
  - clamped on all four edges.
- Two types of loading are considered:
  - uniform load.

  - point load at the center of the plate.

- Owing to practical considerations, only four angles are used: $0°, +45°, -45°$ and $90°$.
- Plate aspect ratios between 1 and 10 are considered.
- Four–node, linear, quadrilateral finite elements are used.
- The mesh consists of 400 nodes.
- The material used is T300/N5208.

where a and b denote the inplane dimensions of the plate.

Weight saving and normalized deflection for thin, clamped, uniform loaded plates are shown in Figure 7.4–3. As we can see, the weight saving is a function of the aspect ratio, the maximum value being 48% for $b/a = 2$. The optimum sublaminate also varies in function of the aspect ratio:

- [0/90] for b/a between 1 and 1.5
- $[0_5/90_2]$ for b/a between 1.5 and 4.4
- [0] for b/a higher than 4.4

The quasi–isotropic sublaminate $[0/+45/-45/90]$ presents low weight saving with respect to the rest of the sublaminates, maximum value being 40% for a square plate.

The optimum laminate thickness can be calculated by means of the following expression:

$$t = 2ka\sqrt{\frac{P}{X}} \qquad (7.4-3)$$

where a is the plate length, P is the applied load, X is the uniaxial tensile strength of a ply along the x–axis, and k is a coefficient that varies as a function of x.

The variation of the nondimensional parameter k with the x–axis and the diagonal are shown in Figures 7.4–4 through 7.4–6.

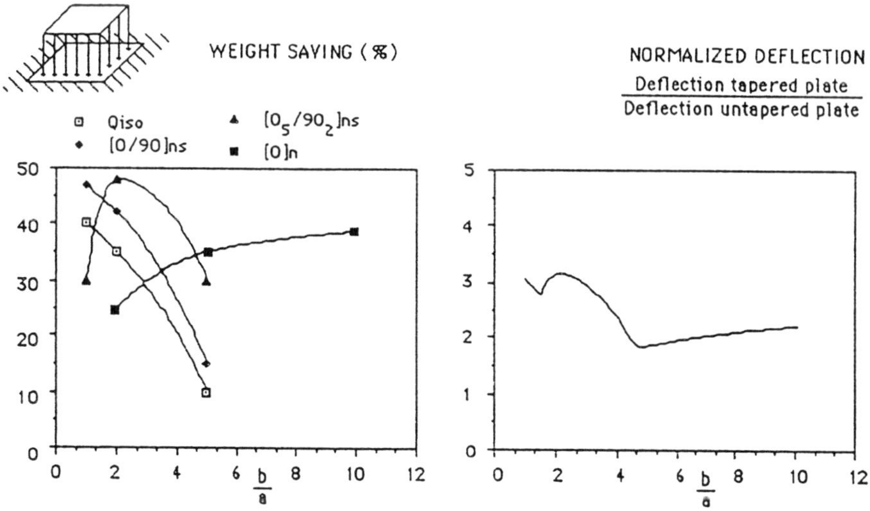

Figure 7.4–3: Weight saving and normalized deflection for a thin, clamped plate subjected to uniformly distributed transverse load.

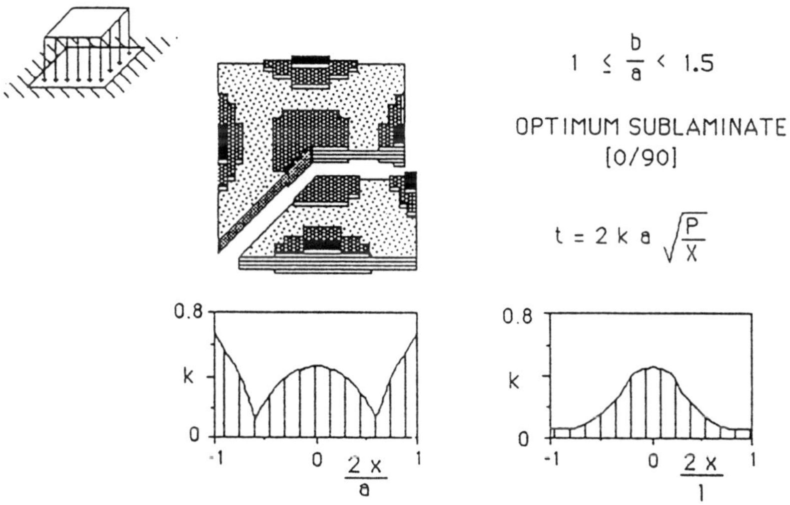

Figure 7.4–4: Representation of an optimal, thin, clamped plate subjected to a uniform distributed transverse load for aspect ratios between 1 and 1.5.

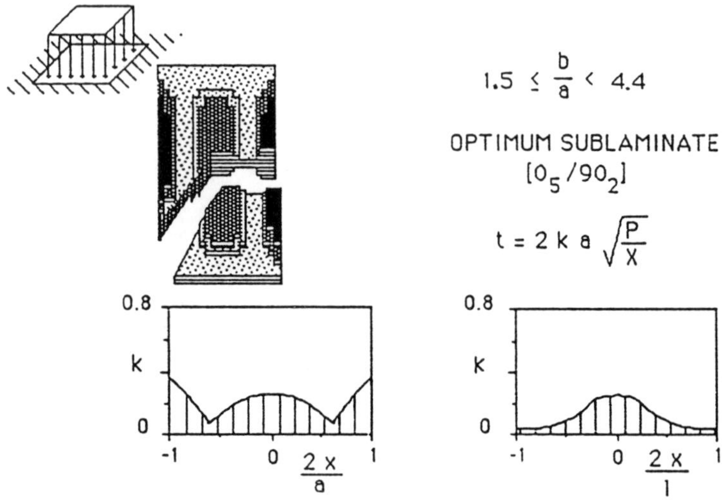

Figure 7.4–5: Representation of an optimal, thin, clamped plate subjected to a uniform distributed transverse load for aspect ratios between 1.5 and 4.4.

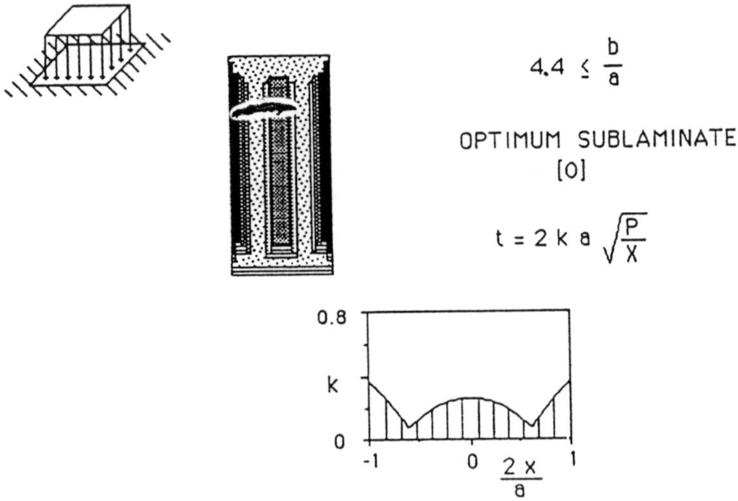

Figure 7.4–6: Representation of an optimal, thin, clamped plate subjected to a uniform distributed transverse load for aspect ratios higher than 4.4.

Weight savings and normalized deflection for thick, clamped, uniform loaded plates are shown in Figure 7.4–7. As we can see, the weight saving is a function of the aspect ratio, the maximum value being 46% for $b/a = 10$. The optimum sublaminate also varies as a function of the aspect ratio:

- $[+45/-45]$ for b/a between 1 and 3.5
- $[0]$ for b/a higher than 3.5

The quasi–isotropic sublaminate $[0/+45/-45/90]$ presents low weight saving with respect to the rest of the sublaminates, the maximum value being 20% for $b/a = 10$.

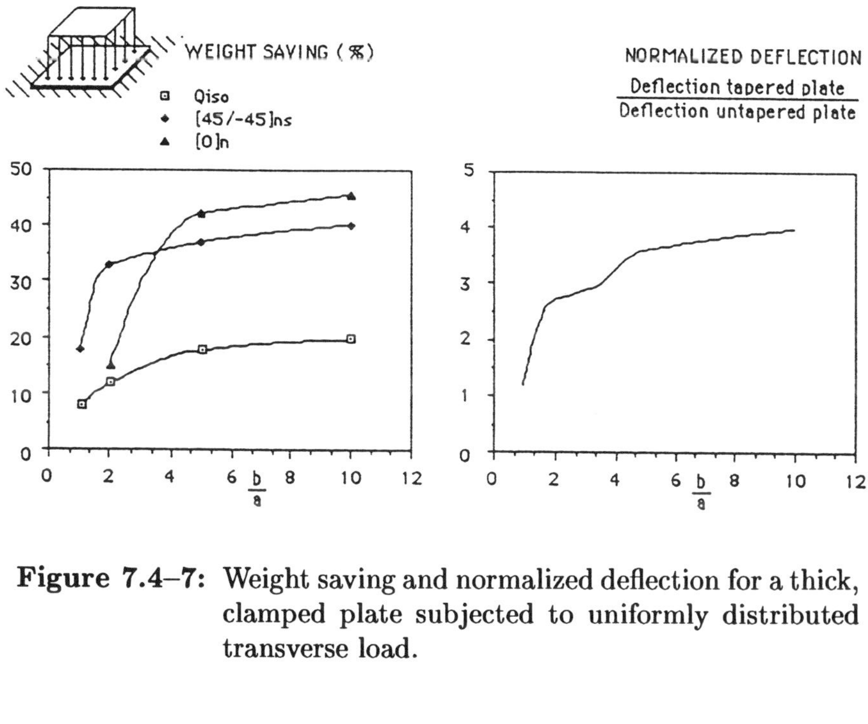

Figure 7.4–7: Weight saving and normalized deflection for a thick, clamped plate subjected to uniformly distributed transverse load.

The optimum laminate thickness can be calculated by means of the following expression:

$$t = 2ka\frac{P}{S} \qquad (7.4-4)$$

The values of the nondimensional parameter k along the $x-$axis and the diagonal are presented in Figures 7.4–8 and 7.4–9.

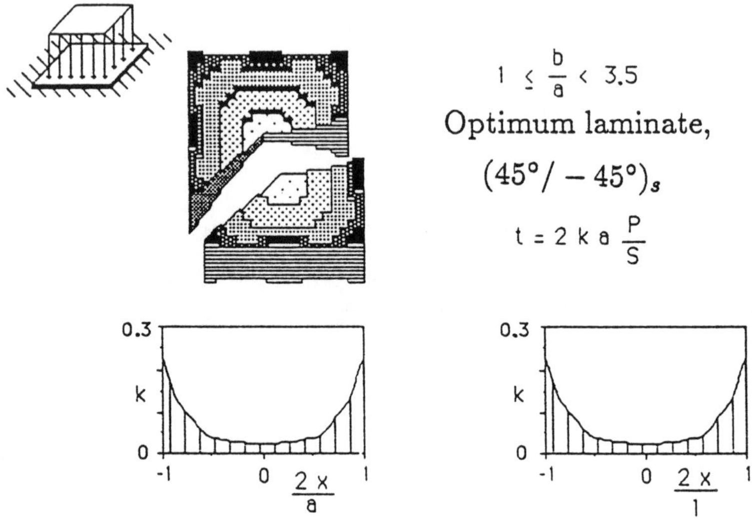

Figure 7.4–8: Representation of an optimal, thick, clamped plate subjected to a uniform distributed transverse load for aspect ratios between 1 and 3.5.

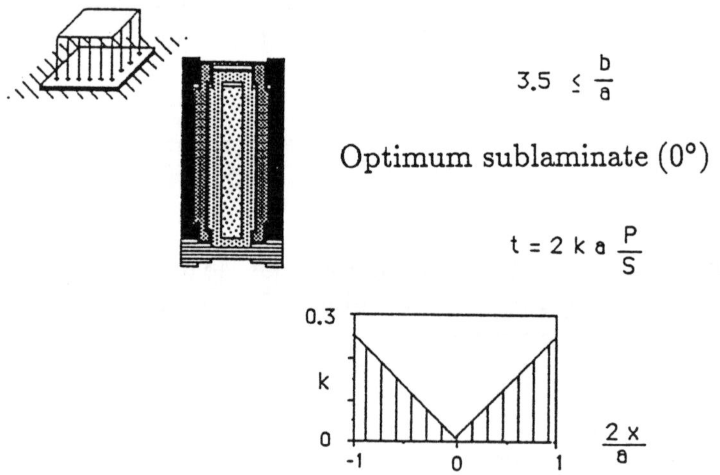

Figure 7.4–9: Representation of an optimal, thick, clamped plate subjected to a uniform distributed transverse load for aspect ratios higher than 3.5.

7.5 CORRELATION OF THEORY AND EXPERIMENT

An experimental study has been carried out in order to assess the accuracy of the theoretical model, as well as to find out the influence of the angle of variation of thickness on the strength of a composite plate. The following information was used:

- AS4/3501–6 Graphite/Epoxy material.
- The three–point static bend test was used.
- The load rate was 0.06 in/min (0.001524 m/min).
- Three types of specimens were tested. Maximum and minimum thickness and span were the same for the three specimens. The only variable parameter was the angle of variation of thickness:
    - Specimen OLCP_7001. Angle of variation of thickness: 6°
    - Specimen OLCP_7002. Angle of variation of thickness: 45°
    - Specimen OLCP_7003. Angle of variation of thickness: 90°
- The span was 8 inches (0.2032 meters).
- The maximum thickness was at the center of the beam: 0.66 inches (0.0167 meters).
- The minimum thickness was at the end of the beam: 0.28 inches (0.007 meters).
- Longitudinal and interlaminar strains were measured using strain gauges. The location of the gauges is reported in Table 7.5–1. The selection of the points was based on the theoretical analysis. Strain gauges were disposed at those points whose strain values were critical, according to the results from the finite element analysis of variable thickness laminated composite plates.

According to the results of the study of failure mechanisms, critical strain components are the longitudinal strain ε_1 and the interlaminar shear strain ε_5. Therefore, both components are analyzed for the three types of specimens described above.

Table 7.5–1: Coordinates x and z of the strain gauges. Values are expressed in inches (meters).

Point	x		z	
1	0.0	(0.0000)	0.100	(0.00254)
2	0.0	(0.0000)	0.200	(0.00508)
3	1.0	(0.0254)	0.350	(0.00889)
4	2.0	(0.0508)	0.415	(0.01054)
5	3.0	(0.0762)	0.600	(0.01524)
6	4.0	(0.1016)	0.510	(0.01295)
7	4.0	(0.1016)	0.350	(0.00889)
8	4.0	(0.1016)	0.000	(0.00000)

The scheme of the specimen OLCP_7001 and its measurements are reported in Figure 7.5–1. This specimen was made by tapering the plate in its inner part. The angle of change of thickness is 6°. The failure load was 4950 lb (22000N). The upper surface does not show any failure, in spite of being tapered. Instead, a crack appeared near the middle plane at the end of the plate where interlaminar shear strains are maximum. The reason for this is that the angle of variation of thickness is very low. The locations of the gauges are shown in Figure 7.5–2. The numerical results of the specimen OLCP_7001 are presented in Tables 7.5–2 through 7.5–4. Tables 7.5–2 and 7.5–3 show longitudinal and interlaminar shear strains, respectively, at the critical points for the failure load. The maximum value is registered at Point 1 ($\varepsilon_5 = 0.0131$).

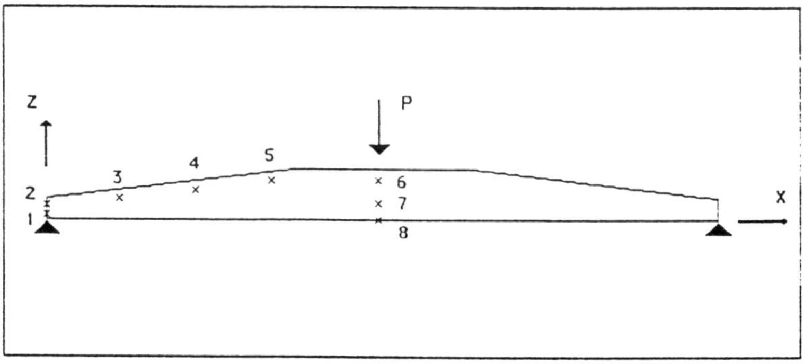

Figure 7.5–1: Measurements of specimen OLCP_7001 in inches (meters).

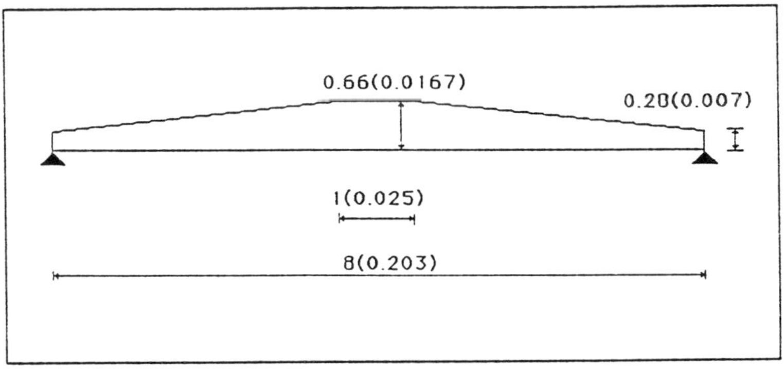

Figure 7.5–2: Representation of specimen OLCP_7001 and position of gauges.

Table 7.5–2: Longitudinal strains ($\varepsilon_1 \times 10^{-3}$) in specimen OLCP_7001. Load: 4950 lb (22,000 N).

Point	Experiment	Theory
8	8.9	8.6

Table 7.5–3: Interlaminar shear strains ($\varepsilon_5 \times 10^{-3}$) in specimen OLCP_7001. Load 4950 lb (22,000 N).

Point	Experiment	Theory
1	13.1*	12.5
2	9.0	8.1
3	6.3	7.3
4	5.7	6.7
5	4.2	5.8
6	3.7	3.1
7	12.7	12.1

*Failure

The maximum vertical displacement is reported in Table 7.5–4. An excellent agreement is found between the experimental data and the theoretical prediction of strains as well as displacements.

Table 7.5–4: Maximum vertical displacement in specimen OLCP_7001. Load: 4950 lb (22,000 N).

Point	Experiment	Theory
8	−0.173 (0.00439)	−0.161 (0.00409)

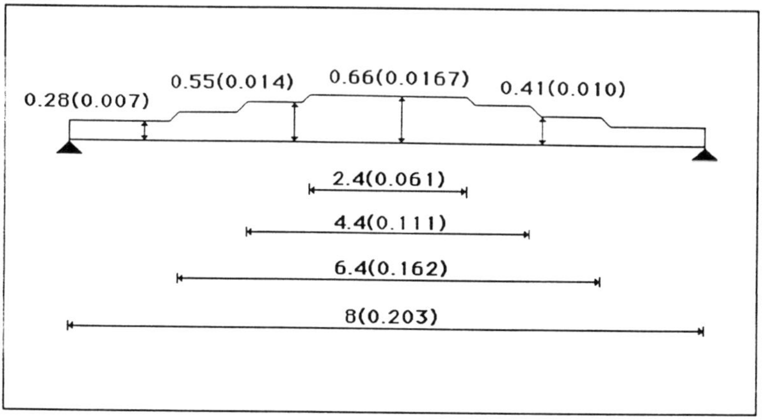

Figure 7.5–3: Measurements of specimen OLCP_7002 in inches (meters).

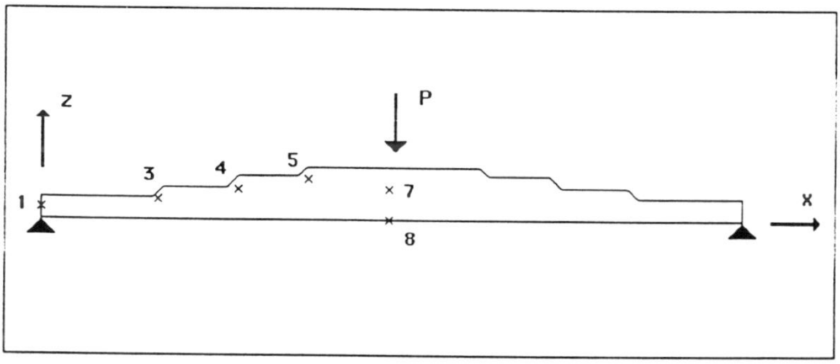

Figure 7.5–4: Representation of specimen OLCP_7002 and position of gauges.

The scheme of the specimen OLCP_7002 and its measurements are reported in Figure 7.5–3. The angle of change of thickness is 45°. Six steps were designed along the specimen. The failure load was 2960 lb (13,150 N). Owing to the fact that the angle of variation of thickness is high, failure appears at the upper surface in the thinnest area of change of thickness. Failure mechanism is an interlaminar shear mode. According to the theoretical results, there is a peak of interlaminar shear strains at the places where the thickness changes. The locations of the gauges are depicted in Figure 7.5–4. The numerical results of the specimen OLCP_7002 are presented in Tables 7.5–5 through 7.5–7. Tables 7.5–5 and 7.5–6 show longitudinal and interlaminar shear strains, respectively, at the critical points for the failure load. The maximum value is registered at Point 3 ($\varepsilon_5 = 0.0131$).

Table 7.5–5: Longitudinal strains ($\varepsilon_5 = 0.0131$) in specimen OLCP_7002. Load: 2960 lb (13,150 N).

Point	Experiment	Theory
8	4.4	3.96

Table 7.5–6: Interlaminar shear strains ($\varepsilon_5 \times 10^{-3}$) in specimen OLCP_7002. Load: 2960 lb (13,150 N).

Point	Experiment	Theory
1	6.0	5.7
3	12.1*	12.5
4	11.0	11.6
5	10.0	9.8
7	2.8	2.3

*Failure

The maximum vertical displacement is reported in Table 7.5–7. For both strains and displacements, an excellent agreement is found between the experiment and the prediction.

Table 7.5–7: Maximum transverse displacement in specimen OLCP_7002. Load: 2960 lb (13,150 N).

Point	Experiment	Theory
8	-0.157 (0.004)	-0.141 (0.0036)

Table 7.5–8: Longitudinal strains ($\varepsilon_1 \times 10^{-3}$) in specimen OLCP_7003. Load: 2485 lb (11,046 N).

Point	Experiment	Theory
8	5.13	4.6

The scheme of the specimen OLCP_7003 and its measurements are reported in Figure 7.5–5. The angle of change of thickness is 90°. Six steps were designed along the specimen. The failure load was 2485 lb (11046 N). Owing to the fact that the angle of variation of thickness is very high, failure appears at the upper surface in the thinnest area of change of thickness at a low failure load. As was the case with specimen OLCP_7002, the failure mechanism is a kind of interlaminar shear mode. As predicted by the analysis, there is a peak of interlaminar shear strains in places where the thickness changes. The locations of the gauges are depicted in Figure 7.5–6. The numerical results of the specimen OLCP_7003 are presented in Tables 7.5–8 through 7.5–10. Tables 7.5–8 and 7.5–9 show longitudinal and interlaminar shear strains in the critical points for the failure load. The maximum value is registered at Point 3 ($\varepsilon_5 = 0.0139$).

The maximum vertical displacement is reported in Table 7.5–10. Excellent agreement is found between the experiment and the prediction of both strains and displacements.

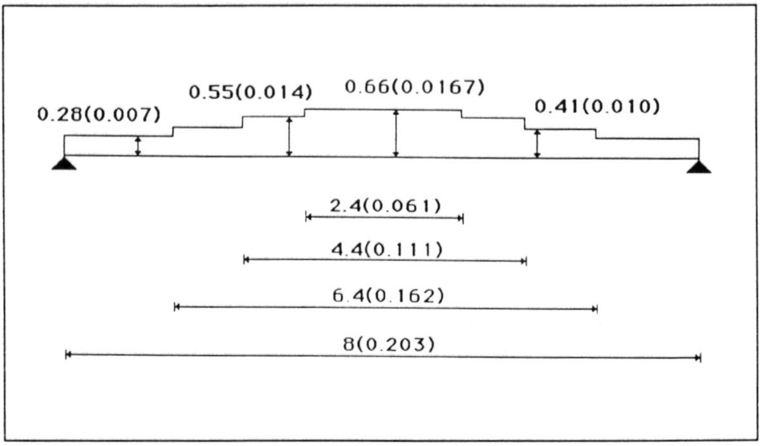

Figure 7.5–5: Measurements of specimen OLCP_7003 in inches (meters).

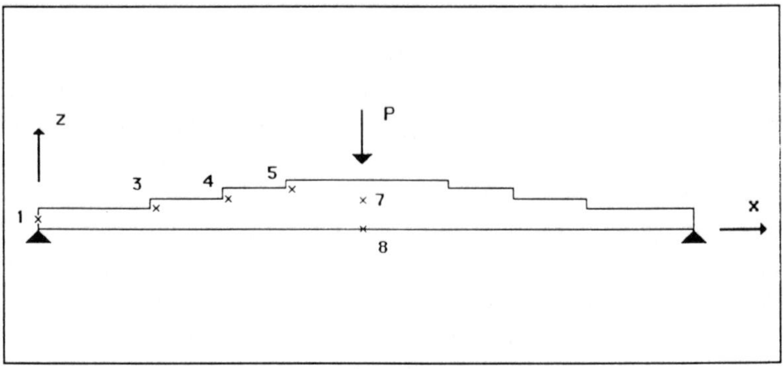

Figure 7.5–6: Representation of specimen OLCP_7003 and position of gauges.

The results of the three specimens tested are presented in Tables 7.5–11 through 7.5–13. These tables show the values of longitudinal and interlaminar shear strains and maximum transverse displacements for a load of 1000 lb (4445 N).

Table 7.5–9: Interlaminar shear strains ($\varepsilon_5 \times 10^{-3}$) in specimen OLCP_7003. Load: 2485 lb (11,046 N).

Point	Experiment	Theory
1	7.25	7.5
3	13.9*	12.5
4	12.18	11.6
5	11.02	9.8
7	3.77	3.4

*Failure

Table 7.5–10: Maximum vertical displacement in specimen OLCP_7003. Load: 2485 lb (11,046 N).

Point	Experiment	Theory
8	−0.171 (0.00432)	−0.157 (0.004)

As expected, the longitudinal strains at the inner surface in the middle of the span are very similar for the three specimens tested. The reason is that the three specimens have the same thickness at the middle of the span and, therefore, the longitudinal strains and stresses should be very close in the three cases.

Table 7.5–12 contains results for the three specimens. On one hand, the maximum values are registered in specimens OLCP_7002 and OLCP_7003, whose angles of variation of thickness have large values (45° and 90°, respectively). On the other hand, in these two cases, the values given by gauges 3, 4, and 5 are critical. This means that the upper surface is the critical one from the failure point of view. On the other hand, specimen OLCP_7001 does not show its maximum value at any of these points, and, therefore, the tapered surface is not a critical place.

Table 7.5–11: Longitudinal strains ($\varepsilon_1 \times 10^{-3}$) in specimens OLCP_7001, OLCP_7002 and OLCP_7003.

Point	OLCP_7001	OLCP_7002	OLCP_7003
8	1.78	1.77	1.76

Table 7.5–12: Interlaminar shear strains ($\varepsilon_5 \times 10^{-3}$) in specimens OLCP_7001, OLCP_7002 and OLCP_7003. Load: 100 lb (4445 N).

Point	OLCP_7001	OLCP_7002	OLCP_7003
1	2.62	2.5	2.4
3	1.26	4.8	5.24
4	1.14	4.2	4.4
5	0.84	3.8	4.0
7	2.54	1.3	1.12

The results reported in Table 7.5–13 point out a higher stiffness of specimen OLCP_7001 with respect to the other two. The smaller the angle of variation of thickness, the better the structure. According to these results, the relation between strains and the angle of variation of thickness is not linear, and the angle of variation of thickness must be very small in order to obtain a better design from the point of view of stiffness and strength.

Table 7.5–13: Maximum transverse displacements in inches (meters) in three specimens. Load: 1000 lb (4445 N).

Point	OLCP_7001	OLCP_7002	OLCP_7003
8	−0.0346 (0.00087)	−0.059 (0.00149)	−0.0628 (0.00159)

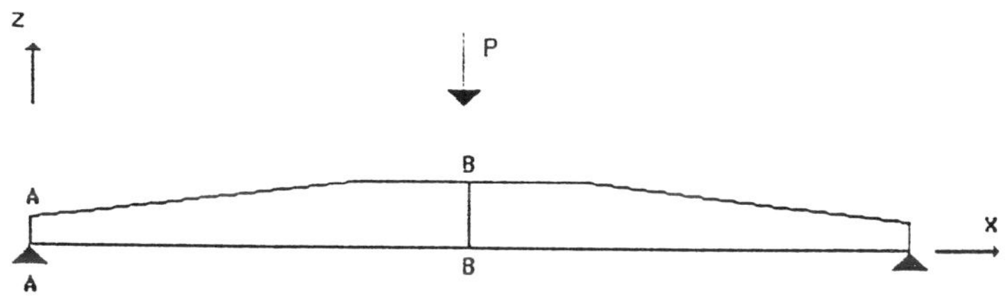

Figure 7.5–7: Representation of Sections AA and BB.

Figures 7.5–8 and 7.5–9 show the distribution of interlaminar strains through the laminate thickness. Both plots refer to the three–point bend test and the specimen OLCP_7001 described above. Figure 7.5–8 shows the theoretical and expermental distributions of interlaminar shear strain ε_5 at section AA. As we can see from Figure 7.5–7, section AA corresponds to the end of the beam. This section is critial because the laminate thickness is minimum there. Figure 7.5–9 shows the distribution of ε_5 at section BB, at the middle of the span (Figure 7.5–8). In both cases, theoretical and experimental distributions are very close.

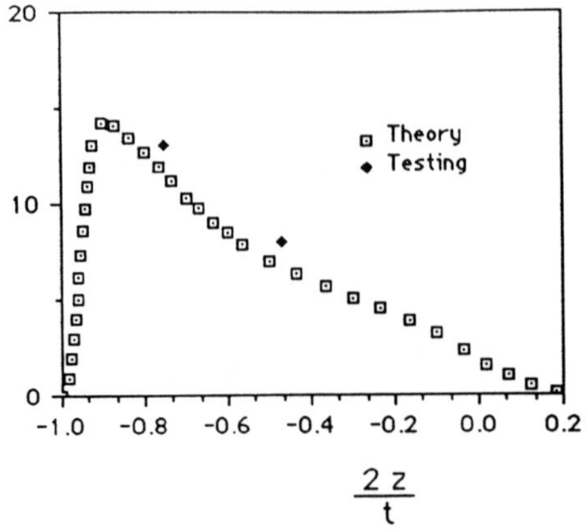

Figure 7.5–8: Distribution of ε_5 through the thickness at Section AA.

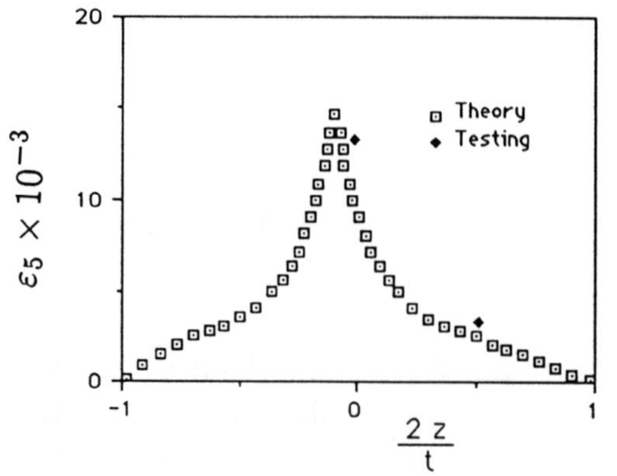

Figure 7.5–9: Distribution of ε_5 through the thickness at Section BB.

Based on the results presented above, the following conclusions can be drawn:

- For small angles of variation of thickness, the strength is very large and the variable thickness effect does not alter the mechanical behavior of the plate. Failure occurs at the middle plane of the plate and corresponds to a delamination mechanism (due to interlaminar shear stress).
- For large values of the angle of variation of thickness, the strength is low, and the variable thickness effect does alter the mechanical behavior of the plate. Failure occurs in the area where the thickness changes, and it corresponds to a delamination mechanism (due to high interlaminar shear stresses provoked by the variable thickness effect).
- The numerical results from theory and experiment are very close for deflections and longitudinal and interlaminar strains.
- Theoretical and experimental distributions of through–thickness strains are also very close (Figures 7.5–8 and 7.5–9).

REFERENCES FOR ADDITIONAL READING

1. Reddy, J. N., "A Simple Higher–Order Theory of Laminated Composite Plates", *J. Appl. Mech.*, **51**, 745, 1984.
2. Miravete, A., "The Finite Element Method Applied to Composites," Esribano, Zaragoza, Spain, 1987.
3. Miravete, A., "Analysis del Comportamiento Resistente de Laminados de Poliester Reforzado y su Aplicacion a los Medios de Transporte," Ph.D. Thesis, University of Zaragoza, Spain, 1984.
4. Miravete, A., "Caractcrisation et mise au Point d'un Element Fini de Grande Precisin Applicable a des Lamines de Polyester Renforce avec Fibres de Verre," *Composites*, **4**, pp. 20, 1986.
5. Miravete, A., "A Simple Finite Element Formulation for 3 Dimensional Laminated Composite Plates," *Proceedings of the Fifth International Conference on Composite Structures (ICCS/5)*, I. H. Marshall (ed.), Vol. 1, pp. 405–417, 1989.

6. Miravete, A., Piedrafita, G., Baselga, S., Bravo, P., "Failure Analysis on Variable Thickness Composite Beams Under 4–Point Test Bending," *Proceedings of the Eighth International Conference on Composite Materials (ICCM/8)*, 36–Q–1 to 36–Q–3, Hawaii, 1991.

7. Miravete, A., "Optimization of Composite Structures," AFWAL/MLBM Technical Report, Dayton, OH, 1989.

8. Miravete, A. and Kim, R., *Failure Modes on Variable Thickness Composite Structures,* SAMPE Series, Vol. 35, Part 2, pp. 1647–1656, Anaheim, CA, 1990.

9. Miravete, A., Kim, R., Peidrafita, G., Baselga, S., "Through Thickness Stress Distribution in Tapered Composite Beams," *Proceedings ECCM4*, pp. 871–876, Stuttgart, 1990.

Chapter Eight

CRASH ANALYSIS

8.1 INTRODUCTION

Nowadays there is a considerable amount of published data on the response of composite tubes to axial crushing. Most of this work is concerned with the use of composites in applications where energy absorption under crash conditions is an important requirement. The work is complementary to a larger body of research on the behavior of metallic and thermoplastic tubes under the same loading conditions. Tubes made from these materials, which are homogeneous, isotropic and ductile, collapse by progressive plastic folding. The fold geometry and the collapse loads depend on the shape and dimensions of the tubes and can be predicted with some accuracy.

For a number of reasons, it is far more difficult to model the axial crushing of tubes made from composite materials. Most structural composites are made from brittle fibers embedded in a polymer matrix, which may be brittle or ductile depending on the choice of the resin. This means that extensive plastic deformation cannot occur, and collapse by plastic folding as one can observe in metal and plastic tubes is impossible. In addition, the properties of composite materials are strongly dependent on the fiber arrangement, the fiber volume fraction and the properties of the fiber-matrix interface, and invariably have some degree of anisotropy.

Two main types of progressive collapse have been reported in tubes made from composite materials. The first resembles the progressive folding mode observed in ductile metal and plastic tubes. This mode has been observed in thin–walled tubes of continuous carbon and glass fiber materials, in a range of tubes of different wall thickness made from tough polymer fibers such as Kevlar™, and in a range of tubes of different wall thickness made from glass fibers and polyester or epoxy resins tested at high temperatures. The second, progressive crushing involves the formation of a zone of microfracture at one end of the tube, which then propagates along the tube at the same speed as the crushing platen.

8.2 THEORETICAL FORMULATION

8.2–1 Introduction

In the linear description of the motion of solid bodies it is assumed that the displacements are infinitely small and that the material is linearly elastic. In addition, it is also assumed that the nature of the boundary conditions remains unchanged during the entire deformation process. These assumptions imply that the displacement vector $\mathbf{u}$ is a linear function of the applied load vector F; *i.e.*, if the displacement under applied load F is $\mathbf{u}$, then the displacement due to a scalar multiple of F, say αF, is $\alpha\mathbf{u}$.

The nonlinearity in solid mechanics arises from two distinct sources. One due to the kinematics of deformation of the body and the other from constitutive behavior (*i.e.*, stress–strain relations). The analyses in which the first type of nonlinearity is considered are called *geometrically* nonlinear analyses, and those in which the second type is considered are called *materially* nonlinear analyses. The geometrically nonlinear analysis can be subclassified according to the type of nonlinearity considered. Two such cases are: (i) large displacements and large rotations but small strains, and (ii) large displacements, large rotations, and large strains. In the first case it is assumed that rotations of line elements are large, but their extensions and changes of angles between two line elements are small. In the second case the extension of a line element and angle changes between two line elements are large, and displacements and rotations of a line element are large as well.

8.2–2 Descriptions of Motion

Consider a deformable body (or a structural system) $\mathcal{B}$ of known geometry, constitution, and loading. Under given geometric restrictions and loading, the body will undergo motion and/or deformation (*i.e.*, geometric changes within the body). If the applied loads are time–dependent, the deformation of the body will be a function of time, *i.e.*, the geometry of the body will change continuously with time. If the loads are applied slowly so that the deformation is only dependent on the loads, the body will take a definitive shape at the end of each load application. Whether the deformation is time dependent or not, the forces acting on the body will always be in equilibrium.

Consider a solid body $\mathcal{B}_0$ at time $t = 0$ (or initially) in which a particle B occupies a position $\mathbf{a}$. Note that B is the name of the particle that occupies the location $\mathbf{a}$ in the reference coordinate system. At time $t > 0$, the body assumes a new shape $\mathcal{B}$, and the particle B occupies the position $\mathbf{x}$ (see Figure 8.2–1).

There are two commonly used descriptions of motion and deformation in continuum mechanics. In the *Lagrangian description*, the motion of a body is referred to a reference configuration $\mathcal{B}_0$. Thus, in the Lagrangian description, the current Cartesian coordinates (x_1, x_2, x_3) of the point $\mathbf{x}$ are expressed in terms of the reference coordinates (a_1, a_2, a_3):

$$x_i = x_i(a_1, a_2, a_3, t) \qquad (8.2-1)$$

For $t = 0$ we have the initial conditions

$$x_i(a_1, a_2, a_3, 0) = a_i \qquad (8.2-2)$$

The displacement vector is then given by

$$\mathbf{u} = \mathbf{x} - \mathbf{a} \quad \text{or} \quad u_i = x_i - a_i \qquad (8.2-3)$$

where u_i denotes the i–th component of the displacement vector.

In the *spatial* or *Eulerian description* of a body, the motion is referred to the current configuration occupied by the body $\mathcal{B}$. The spatial description focusses attention on a given region of space instead of on a given solid body, and this is the description most used to study flow of a material. In the Lagrangian description the coordinate system **X** is fixed on a given solid body in its undeformed configuration, and its position **x** at any time is referred to the material coordinate a_i.

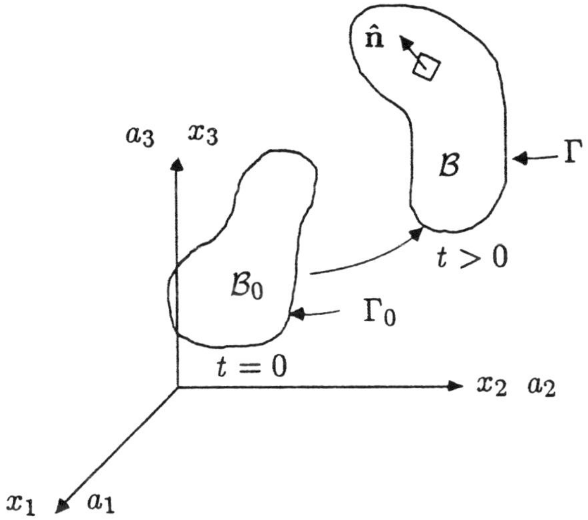

Figure 8.2–1: Reference and deformed configurations of a body.

8.2–3 Governing Equations

The equations of motion (or conservation of linear momentum) in the spatial description can be expressed as

$$\sigma_{ij,j} + \rho f_i = \rho \ddot{x}_i \qquad (8.2-4)$$

where σ_{ij} denote the Cartesian components of the Cauchy stress tensor, ρ is the density of the deformed body, and f_i are the Cartesian components of the body force vector (measured per unit mass). Note that $\ddot{x}_i$ denotes the acceleration. Equation (4) is subjected to stress boundary conditions

$$\sigma_{ij} n_j = t_i(t) \quad \text{on} \quad \Gamma_1 \qquad (8.2-5)$$

the displacement boundary conditions

$$u_i(a_1, a_2, a_3, t) = D_i(t) \quad \text{on} \quad \Gamma_2 \qquad (8.2-6)$$

and the continuity (or contact) conditions

$$(\sigma_{ij}^{+} - \sigma_{ij}^{-}) n_j = 0 \quad \text{on} \quad \Gamma_3 \qquad (8.2-7)$$

where n_i denote the components of a unit normal vector to the boundary $\Gamma = \Gamma_1 + \Gamma_2$ of the body $\mathcal{B}$, and Γ_3 denotes the contact surface.

Conservation of mass leads to

$$\rho J = \rho_0 \qquad (8.2-8)$$

where ρ_0 is the density in the reference configuration, and J is the determinant of the Jacobian (or relative volume)

$$\mathbf{J} = \frac{\partial \mathbf{x}}{\partial \mathbf{a}} \equiv \mathbf{F}\ , \quad J = \det\left(\frac{\partial \mathbf{x}}{\partial \mathbf{a}}\right) \qquad (8.2-9a)$$

and $\mathbf{F}$ denotes the deformation gradient, with components F_{ij}

$$F_{ij} = \frac{\partial x_i}{\partial a_j} \qquad (8.2-9b)$$

The conservation of energy can be expressed as

$$\dot{E} = \rho_0 s_{ij} \dot{\varepsilon}_{ij} - (P + Q)\dot{J} \qquad (8.2-10)$$

where E is the internal energy density, s_{ij} are the deviatoric components of stress, ε_{ij} are the components of the Eulerian strain tensor, P is the pressure, and Q is the bulk viscosity:

$$\varepsilon_{ij} = \frac{1}{2}\left(\frac{\partial u_i}{\partial x_j} + \frac{\partial u_j}{\partial x_i}\right) \qquad (8.2-11a)$$

$$s_{ij} = \sigma_{ij} + (P+Q)\delta_{ij} \quad , \quad P = -\frac{1}{3}\sigma_{kk} - Q \qquad (8.2-11b)$$

In Eqs. (4)–(11), summation convention (*i.e.*, sum on repeated indices over their range) is used, δ_{ij} denotes the Kronecker delta (see [1–3]), and a superposed dot on variables denotes their total time derivative.

The weak form equivalent to Eqs. (4)–(7) is the same as the statement of the principle of virtual displacements over an element $\mathcal{B}_e$ (see Reddy [4])

$$0 = \int_{\mathcal{B}_e} \rho \ddot{x}_i \, \delta u_i \, dv + \int_{\mathcal{B}_e} \sigma_{ij} \, \delta u_{i,j} \, dv - \int_{\mathcal{B}_e} \rho f_i \, \delta u_i \, dv - \int_{\Gamma_e} t_i \, \delta u_i \, ds = 0 \qquad (8.2-12)$$

where δ denotes the variational symbol (*i.e.*, δu_i denotes small virtual variation of u_i).

8.2–4 Spatial Finite Element Discretization

A mesh of trilinear solid hexahedral finite elements is used with the representation

$$u_i(\mathbf{a}, t) = u_i(\mathbf{a}(\xi, \eta, \zeta), t) = \sum_{j=1}^{8} \psi_j(\xi, \eta, \zeta) u_i^j(t) \qquad (8.2-13)$$

where ψ_j are the interpolation (or shape) functions of the *master* eight–node hexahedron element, defined as [5]

$$\psi_j = \frac{1}{8}(1+\xi\xi_j)(1+\eta\eta_j)(1+\zeta\zeta_j) \qquad (8.2-14)$$

where ξ_j, η_j, ζ_j take on the values (± 1) of the coordinates of the element nodes in the natural coordinate system (ξ, η, ζ), and u_i^j denotes the displacement of the j–th node in the i–th direction (see Figure 8.2–2).

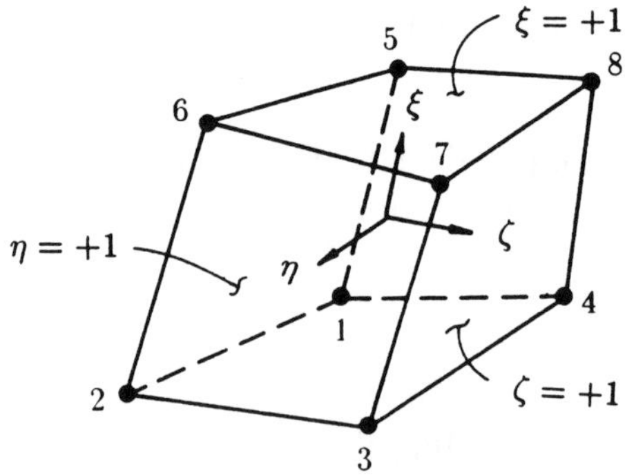

Figure 8.2–2: Trilinear hexahedral finite element.

Substituting Eq. (13) into the statement of the principle of virtual displacements (12) for a typical element $\mathcal{B}_e$, we obtain the finite element model (see Reddy [4], Chapter 12; Zienkiewicz and Taylor [5])

$$\{0\} = \left(\int_{\mathcal{B}_e} \rho [N]^T [N]\, dv\right)\{a\} + \int_{\mathcal{B}_e} [B]^T\{\sigma\}\, dv - \int_{\mathcal{B}_e} \rho [N]^T\{f\}\, dv - \int_{\Gamma_e} [N]^T\{t\} ds \qquad (8.2-15a)$$

or

$$[M]\{a\} = \{R\} \tag{8.2 - 15b}$$

where

$$[M] = \int_{\mathcal{B}_e} \rho [N]^T [N] \, dv \tag{8.2 - 16a}$$

$$\{R\} = \int_{\mathcal{B}_e} \rho [N]^T \{f\} \, dv + \int_{\Gamma_e} [N]^T \{t\} ds - \int_{\mathcal{B}_e} [B]^T \{\sigma\} \, dv \tag{8.2 - 16b}$$

$$[N] = \begin{bmatrix} \psi_1 & 0 & 0 & \psi_2 & 0 & \cdots & \psi_8 & 0 & 0 \\ 0 & \psi_1 & 0 & 0 & \psi_2 & 0 & \cdots & \psi_8 & 0 \\ 0 & 0 & \psi_1 & 0 & 0 & \psi_3 & 0 & \cdots & \psi_8 \end{bmatrix} \tag{8.2 - 17}$$

$$[B] = \begin{bmatrix} \frac{\partial}{\partial x} & 0 & 0 \\ 0 & \frac{\partial}{\partial y} & 0 \\ 0 & 0 & \frac{\partial}{\partial z} \\ \frac{\partial}{\partial y} & \frac{\partial}{\partial x} & 0 \\ 0 & \frac{\partial}{\partial z} & \frac{\partial}{\partial y} \\ \frac{\partial}{\partial z} & 0 & \frac{\partial}{\partial x} \end{bmatrix} [N] \,, \quad \{\varepsilon\} = [B]\{u\} \tag{8.2 - 18}$$

$$\begin{Bmatrix} \ddot{x}_1 \\ \ddot{x}_2 \\ \ddot{x}_3 \end{Bmatrix} = [N] \begin{Bmatrix} a_{x1} \\ a_{y1} \\ a_{z1} \\ \cdot \\ \cdot \\ \cdot \\ a_{x8} \\ a_{y8} \\ a_{z8} \end{Bmatrix} = [N]\{a\} \tag{8.2 - 19}$$

$$\{f\} = \begin{Bmatrix} f_x \\ f_y \\ f_z \end{Bmatrix} \,, \quad \{t\} = \begin{Bmatrix} t_x \\ t_y \\ t_z \end{Bmatrix} \,, \quad \{\sigma\} = \begin{Bmatrix} \sigma_{xx} \\ \sigma_{yy} \\ \sigma_{zz} \\ \sigma_{xy} \\ \sigma_{yz} \\ \sigma_{zx} \end{Bmatrix} \tag{8.2 - 20}$$

and $\{u\}$ represents the nodal displacements vector, and $\{a\}$ represents the nodal accelerations vector.

Equations (15) represent a set of coupled equations. The equations can be uncoupled by diagonalizing the mass matrix $[M]$. The row–sum (see Reddy [4]) criterion can be used to diagonalize the consistent mass matrix defined in Eq. (16) ($m_{ij} = 0$ when $i \neq j$):

$$m_{ii} = \int_{\mathcal{B}_e} \rho\psi_i \sum_{j=1}^{8} \psi_j \, dv = \int_{\mathcal{B}_e} \rho\psi_i \, dv \qquad (8.2-21)$$

The equations (15) are assembled for a given finite element mesh and are solved after applying the boundary conditions. Evaluation of the integrals appearing in Eq. (16) are evaluated using the Gauss quadrature rule (see [4,5]).

8.2–5 Computational Aspects

Approximation of the geometry

Note that the integrals in Eq. (16) involve derivatives of the shape functions with respect to the global coordinates x_i ($x_1 = x, x_2 = y, x_3 = z$). Since the shape functions ψ_i are functions of the normalized (local) coordinates (ξ, η, ζ), we must introduce a transformation from (x, y, z) to (ξ, η, ζ) in order only to evaluate the integrals (*i.e.*, the integrals are evaluated on the master element; see Reddy [4] for details). Using the isoparametric formulation, where the geometry as well as the dependent variables are approximated using the same interpolation functions, we write

$$x_i(\mathbf{a}, t) = \sum_{j=1}^{8} \psi_j(\xi, \eta, \zeta) x_i^j(t) \qquad (8.2-22)$$

where x_i^j denotes the coordinate of the j–th node in the i–th direction. Then by the chain rule of differentiation, we have

$$\begin{aligned}\frac{\partial\psi_i}{\partial\xi} &= \frac{\partial\psi_i}{\partial x}\frac{\partial x}{\partial\xi} + \frac{\partial\psi_i}{\partial y}\frac{\partial y}{\partial\xi} + \frac{\partial\psi_i}{\partial z}\frac{\partial z}{\partial\xi} \\ \frac{\partial\psi_i}{\partial\eta} &= \frac{\partial\psi_i}{\partial x}\frac{\partial x}{\partial\eta} + \frac{\partial\psi_i}{\partial y}\frac{\partial y}{\partial\eta} + \frac{\partial\psi_i}{\partial z}\frac{\partial z}{\partial\eta} \\ \frac{\partial\psi_i}{\partial\zeta} &= \frac{\partial\psi_i}{\partial x}\frac{\partial x}{\partial\zeta} + \frac{\partial\psi_i}{\partial y}\frac{\partial y}{\partial\zeta} + \frac{\partial\psi_i}{\partial z}\frac{\partial z}{\partial\zeta}\end{aligned} \qquad (8.2-23a)$$

or, in matrix form

$$\left\{\begin{matrix}\frac{\partial\psi_i}{\partial\xi}\\ \\ \frac{\partial\psi_i}{\partial\eta}\\ \\ \frac{\partial\psi_i}{\partial\zeta}\end{matrix}\right\} = \begin{bmatrix}\frac{\partial x}{\partial xi} & \frac{\partial y}{\partial\xi} & \frac{\partial z}{\partial\xi}\\ \\ \frac{\partial x}{\partial\eta} & \frac{\partial y}{\partial\eta} & \frac{\partial z}{\partial\eta}\\ \\ \frac{\partial x}{\partial\zeta} & \frac{\partial y}{\partial\zeta} & \frac{\partial z}{\partial\zeta}\end{bmatrix}\left\{\begin{matrix}\frac{\partial\psi_i}{\partial x}\\ \\ \frac{\partial\psi_i}{\partial y}\\ \\ \frac{\partial\psi_i}{\partial z}\end{matrix}\right\} = \mathcal{J}\left\{\begin{matrix}\frac{\partial\psi_i}{\partial x}\\ \\ \frac{\partial\psi_i}{\partial y}\\ \\ \frac{\partial\psi_i}{\partial z}\end{matrix}\right\} \qquad (8.2-23b)$$

by inverting the Jacobian matrix (of the transformation), **J**, the global derivatives of the shape functions can be expressed in terms of the known derivatives of the shape functions with respect to the local coordinates:

$$\left\{\begin{matrix}\frac{\partial\psi_i}{\partial x}\\ \\ \frac{\partial\psi_i}{\partial y}\\ \\ \frac{\partial\psi_i}{\partial z}\end{matrix}\right\} = \mathcal{J}^{-1}\left\{\begin{matrix}\frac{\partial\psi_i}{\partial\xi}\\ \\ \frac{\partial\psi_i}{\partial\eta}\\ \\ \frac{\partial\psi_i}{\partial\zeta}\end{matrix}\right\} \qquad (8.2-24)$$

Volume integration

The volume integration is carried out by means of the Gauss quadrature. For instance, if g is a function of $\mathbf{x}$ defined over the element volume $\mathcal{B}_e$, and N is the number of integration points, then the expression

$$\int_{\mathcal{B}_e} g\ dv = \int_{-1}^{1}\int_{-1}^{1}\int_{-1}^{1} g|\mathcal{J}|\ d\xi d\eta d\zeta \qquad (8.2-25)$$

is approximated by

$$\int_{\mathcal{B}_e} g \; dv \simeq \sum_{j=1}^{N}\sum_{k=1}^{N}\sum_{\ell=1}^{N} G(\xi_i, \eta_k, \zeta_\ell) \; W_j W_k W_\ell \qquad (8.2-26)$$

where W_i are the Gauss weights, and $(\xi_i, \eta_k, \zeta_\ell)$ are the Gauss point coordinates (in the master element), and $G = g\mathcal{J}$.

Equation (26) gives the exact value of the integral whenever the integrand G is a polynomial, and the number of Gauss points is equal to [4]

$$N = \left[\frac{p+1}{2}\right] \qquad (8.2-27)$$

where p is polynomial degree of the integrand G, and $[\cdot]$ donotes the nearest integer to the quotient $(p+1)/2$. Thus a linear polynomial (in each of the coordinates) is exactly integrated by the one–point Gauss rule, and a quadratic polynomial is exactly evaluated by the $2 \times 2 \times 2$ Gauss rule.

Surface integration

When the boundary tractions $\mathbf{t}(t_i)$ are prescribed (*i.e.*, known) on the boundary, then the surface of the element representing that boundary will have known tractions, which must be evaluated as per the surface integral (the second term) in Eq. (16). Of course, when the traction is not prescribed on the boundary, the corresponding displacements are expected to be prescribed.

Consider a surface segment of a typical hexahedral (not the master) element of the mesh. Since the surface is a two–dimensional domain, it can be seen as the two–dimensional quadrilateral element in space (*i.e.*, the specification of its geometry requires three coordinates). In order to locate any points on the surface, we define the position vector r as (see Figure 8.2–3)

$$\mathbf{r} = f_1(\xi, \eta)\mathbf{i}_1 + f_2(\xi, \eta)\mathbf{i}_2 + f_3(\xi, \eta)\mathbf{i}_3 \qquad (8.2-28)$$

where $\mathbf{i}_i$ denotes the Cartesian basis vector associated with x_i–direction, and

$$f_i(\xi,\eta) = \sum_{j=1}^{4} \psi_j x_i^j \tag{8.2 – 29}$$

and ψ_j denote the bilinear interpolation functions

$$\psi_j = \frac{1}{4}(1+\xi\xi_j)(1+\eta\eta_j) \tag{8.2 – 30}$$

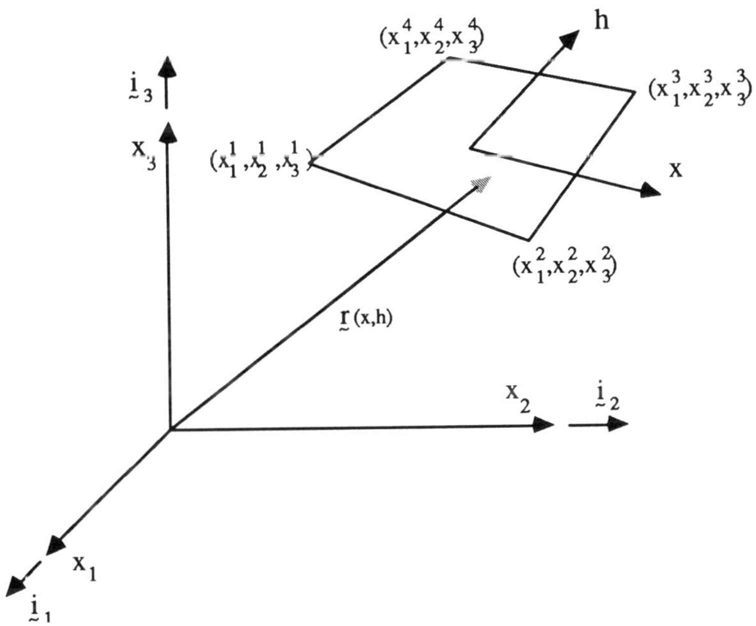

Figure 8.2–3: Parametric representation of a surface segment.

The differential surface element ds can be written in curvilinear coordinates as

$$ds = \mathcal{J}_2 \; d\xi d\eta \tag{8.2 – 31}$$

where $\mathcal{J}_2$ is the surface Jacobian defined by

$$\mathcal{J}_2 = \left| \frac{\partial \mathbf{r}}{\partial \xi} \times \frac{\partial \mathbf{r}}{\partial n} \right| \qquad (8.2-32)$$

The normal vector **n** to the surface segment is

$$\mathbf{n} = \mathcal{J}_2^{-1} \left(\frac{\partial \mathbf{r}}{\partial \xi} \times \frac{\partial \mathbf{r}}{\partial \eta} \right) \qquad (8.2-33)$$

The global components t_i of the traction vector **t** can be written as

$$t_i = \sum_{j=1}^{4} \psi_j T_i^j \qquad (8.2-34)$$

where T_i^j is the value of t_i at the $j-$th node.

Then the surface integral for an element in Eq. (16) is evaluated by adding up the contributions from six integrals corresponding to the six faces of a hexahedral element, where the integral for each surface is given by

$$\int_{-1}^{1} \int_{-1}^{1} \mathbf{N}^T \mathbf{t} \; \mathcal{J}_2 \; d\xi d\eta \qquad (8.2-35)$$

This integral is evaluated numerically using the Gauss rule in two dimensions.

8.3 CORRELATION OF THEORY AND EXPERIMENT

With the aim of verifying the theoretical (*i.e.*, finite element) model used, experimental data for some tubes was obtained. The number of parameters influencing the final results is very large. The errors would have higher absolute values than those introduced in static analyses because of the complexity of the calculations and the simplifications introduced in the dynamic model contribute to the error.

8.3–1 Pultruded Tubes

The pultruded tubes considered were made of E–glass fiber and polyester resin, with an upper face of middle mat and the rest of the tube thickness with fibers placed in the longitudinal direction of the tube. In the model development, different properties in the upper face and the rest of the tube thickness, considered as those of a unidirectional laminate, were introduced.

In the first case, the model consisted of a tube with the end machined in the inner part with a 45° triggering device over half the thickness of the tube wall. In the second case, the model consisted of a tube with the end machined in the inner part with a 45° degree triggering device over the total thickness of the tube wall.

Dimensions of the specimen

Specimen 1:

• Inner diameter	27 mm
• External diameter	32 mm
• Wall thickness	3 mm
• Tube length	70 mm

Specimen 2:

• Inner diameter	22 mm
• External diameter	27 mm
• Wall thickness	2.5 mm
• Tube length	70 mm

Characteristics of the material

• Density	1.690 gr/cm^3
• % fiber weight	48
• % fiber volume	31
• Mat ply thickness	0.4 mm
• Unidirectional zone thickness	2.1 mm

The following data for the elastic stiffness properties and material strengths were obtained when characterizing the material in its principal material directions:

E_x=13830 MPa,	X=94 MPa
E_y=3640 MPa,	X'=78 MPa
E_z=3640 MPa,	Y=60 MPa
v_{xy}=0.20,	Y'=79 MPa
v_{xy}=0.20,	Z=60 MPa
v_{yz}=0.33,	Z'=79 MPa
G_{xy}=2320 MPa,	S_{xy}=40 MPa
G_{xz}=2320 MPa,	S_{xz}=24 MPa
G_{yz}=1100 MPa,	S_{yz}=24 MPa

The theoretical and experimental comparisons for each of the correlated tubes are shown in Figures 8.3–1 through 8.3–4 (also see Tables 8.3–1 and 8.3–2). From these figures we note that, after the initial period of application of the load, the crushing is stable with an increase in the load.

Table 8.3–1: Comparison of experimental data with the theoretical results for pultruded tubes.

	Experiment	Theory	% Error
Peak load (N)	48,010	43,130	10.1
Maintained average load (N)	23,400	20,230	12.2
Absorbed specific energy (J/gr)	26.83	23.61	12.1

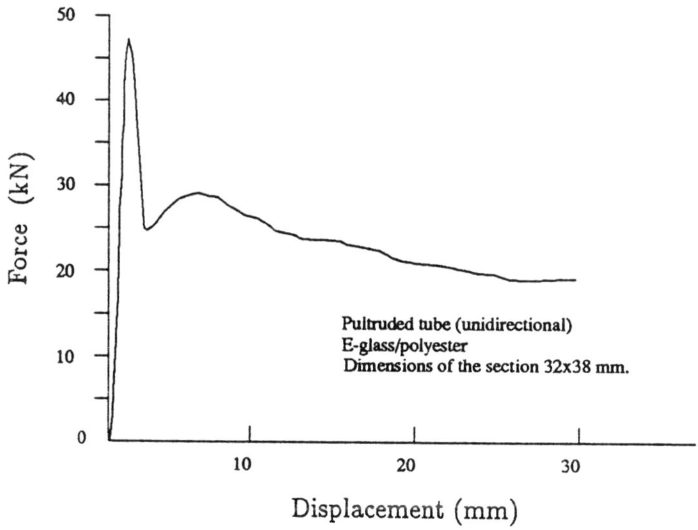

Figure 8.3–1: Force–displacement plot for a triggering device at the half of the inner thickness (experiment).

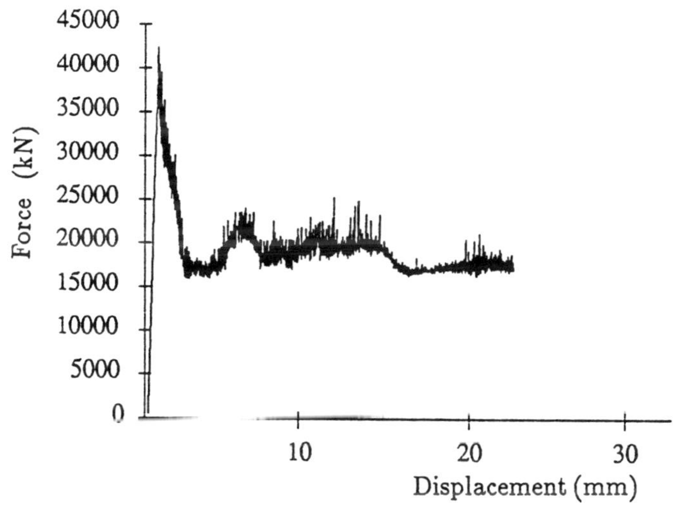

Figure 8.3–2: Force–displacement plot for a triggering device at the half of the inner thickness (theory).

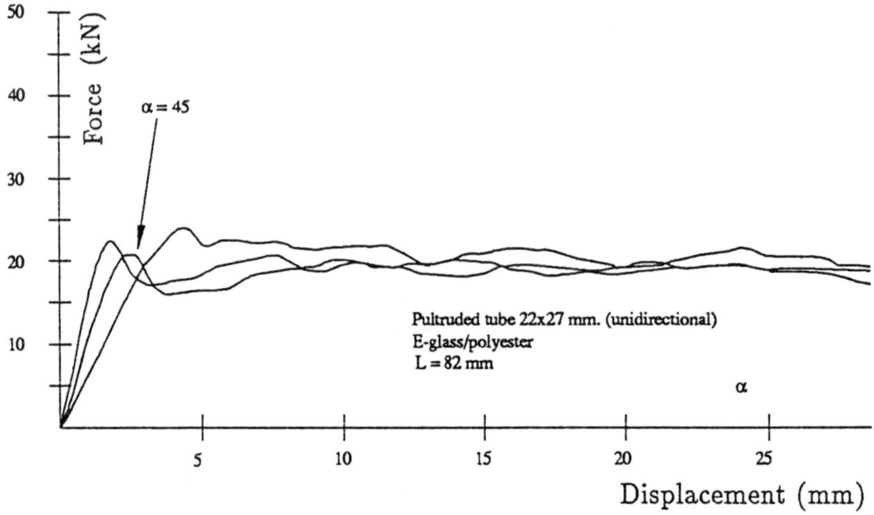

Figure 8.3–3: Force–displacement plot for inner triggering device (experiment).

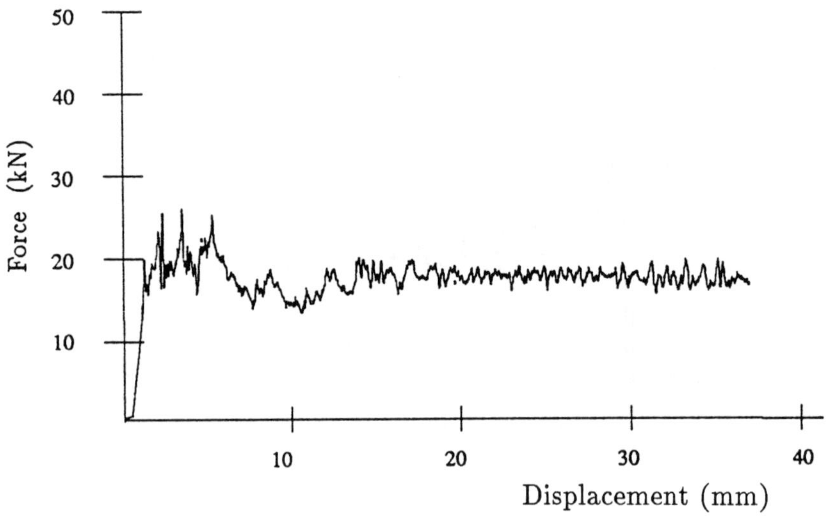

Figure 8.3–4: Force–displacement plot for inner triggering device (theory).

Table 8.3–2: Comparison of experimental data with the theoretical results for pultruded tubes.

	Experiment	Theory	% Error
Peak load (N)	21,160	26,240	24
Maintained average load (N)	18,320	16,250	11.3
Absorbed specific energy (J/gr)	67	59.3	11.4

8.3–2 The 3–D Fabrics

Correlations of three different tests are discussed in this section. The first correlation concerns a 3–D sandwich structure reinforced with carbon plies. Figures 8.3–5 and 8.3–6 show the plots of force versus displacement of the first test, where we note the similarity between the test results and analytical predictions. There is an increase in the load at about 12 mm of displacement due to the fact that 3–D skins are in contact with the wedge, and the load decreases once these skins have been bent.

Figures 8.3–7 and 8.3–8 show force–displacement plots of the theory and experiment of the second test, where we can observe good agreement. In this case, the average level of load in the numerical simulation is lower than that of the test. From Figure 8.3–9, we can see how the skins, due to the wedge shape, slide on it, increasing the level of load in a progressive way. Figures 8.3–10 and 8.3–11 show the force–displacement plots of the third correlation. Figure 8.3–12, corresponding to the third simulation, shows how the skins are in contact with the wedge with a high incidence angle, thus a sharp peak is obtained in the range of displacements (15–20 mm).

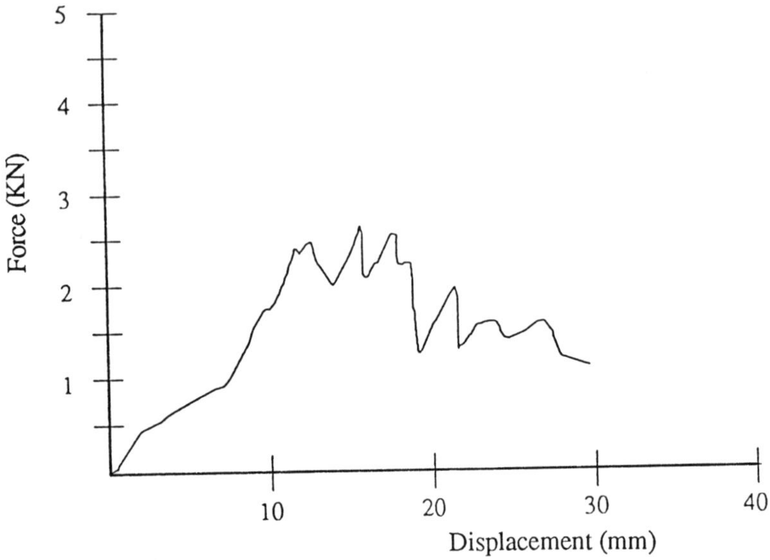

Figure 8.3–5: Force–displacement plot for 3–D fabrics (experiment; Test 1).

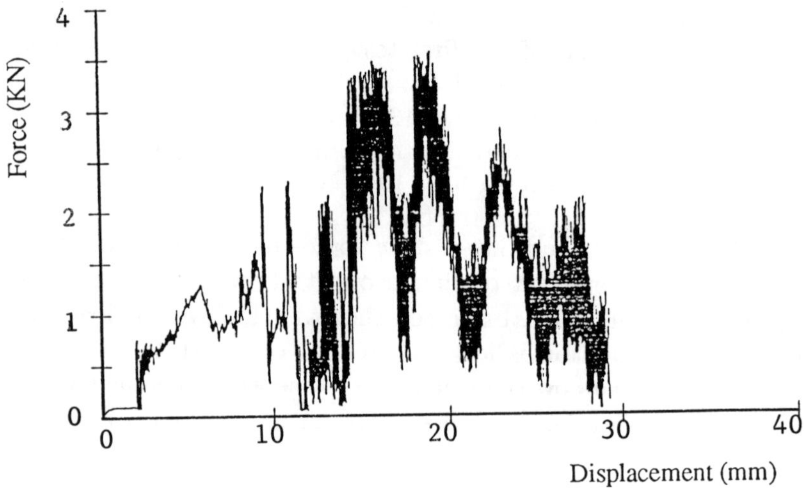

Figure 8.3–6: Force–displacement plot for 3–D fabrics (theory; Test 1).

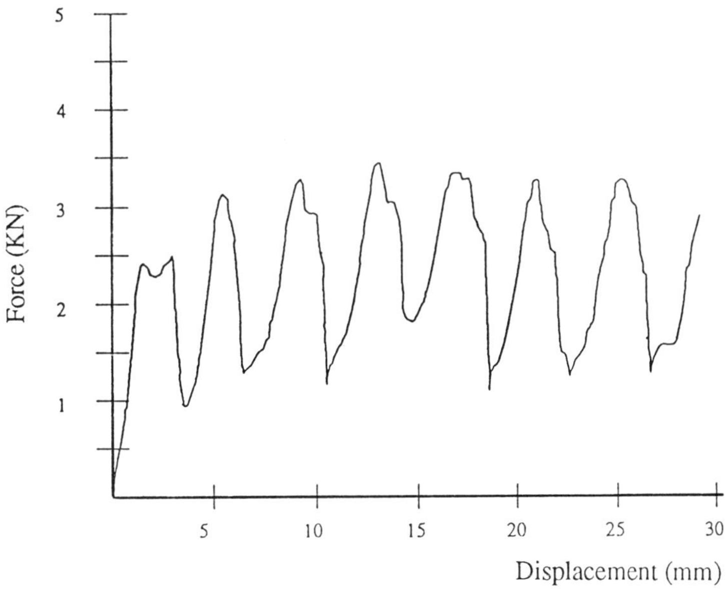

Figure 8.3–7: Force–displacement plot for 3–D fabrics (experiment; Test 2).

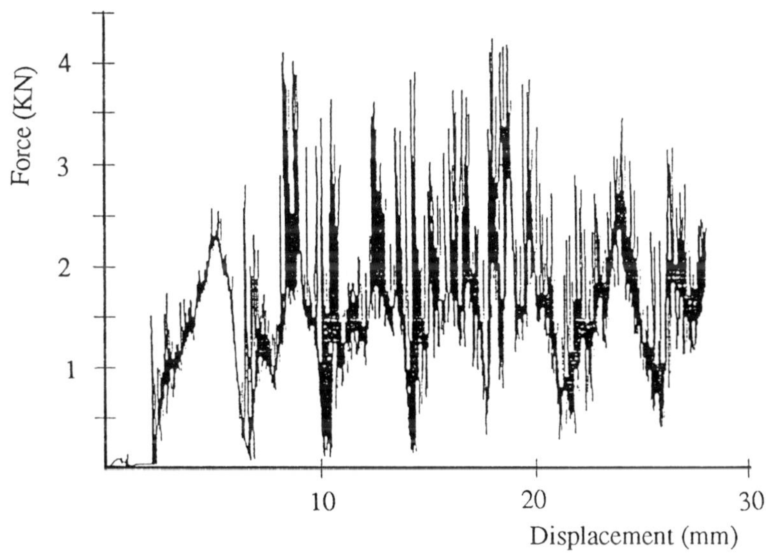

Figure 8.3–8: Force–displacement plot for 3–D fabrics (theory; Test 2).

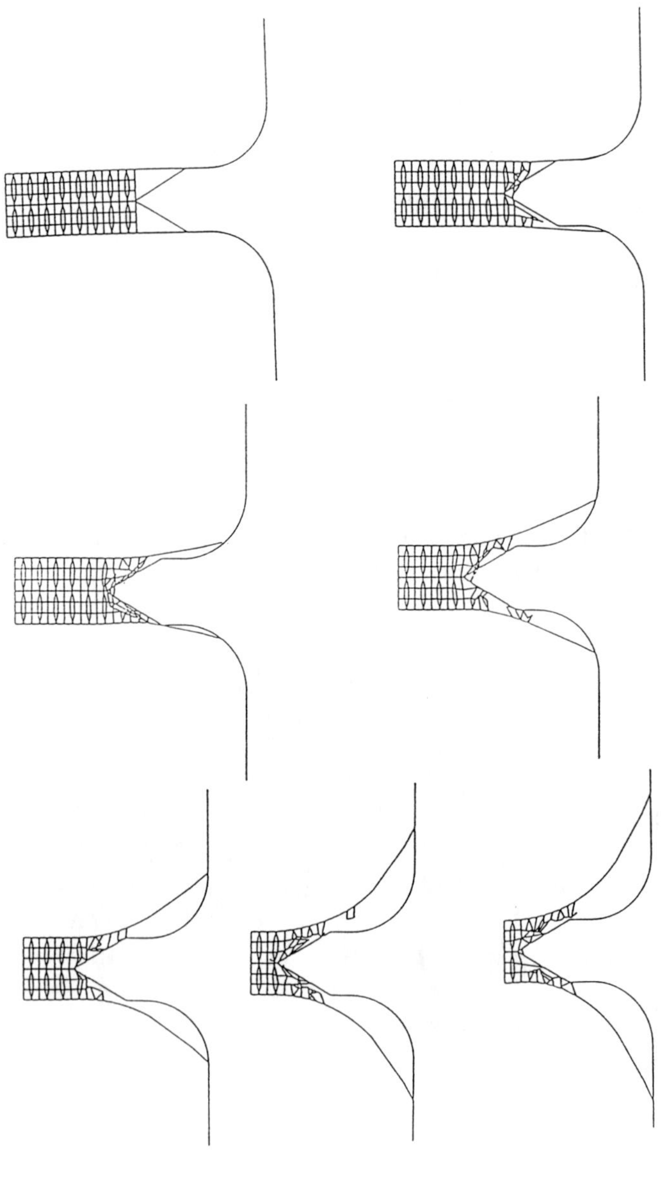

Figure 8.3–9: Sliding of skins due to wedge shape.

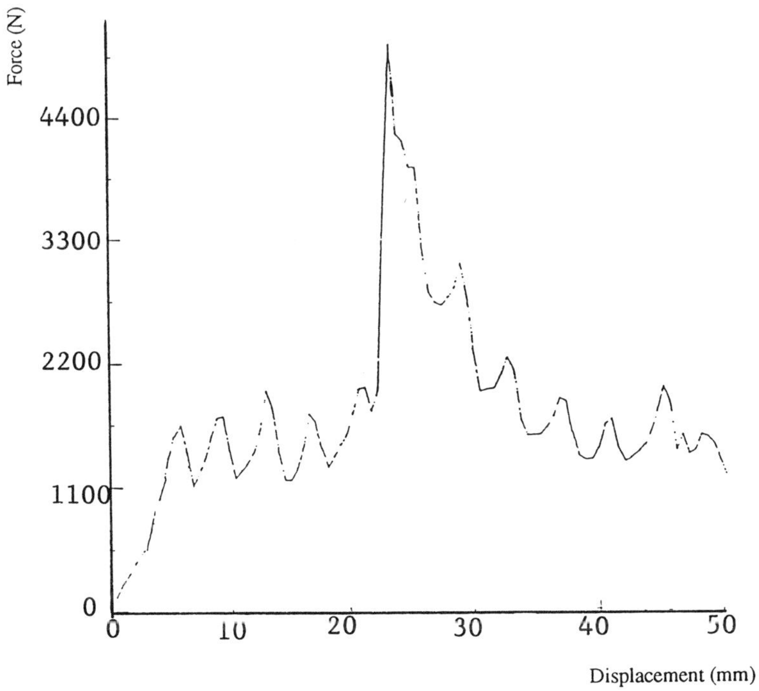

Figure 8.3–10: Plot of force versus displacement for 3–D fabrics (experiment; Test 3).

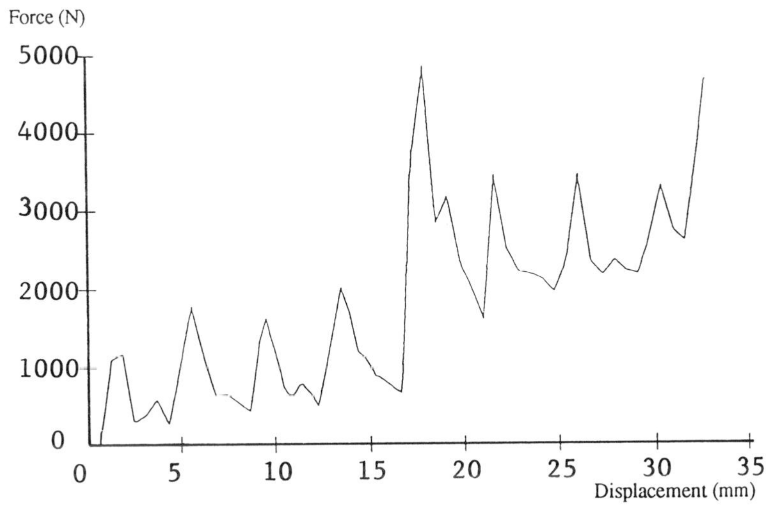

Figure 8.3–11: Force–displacement plot for inner triggering device (theory; Test 3).

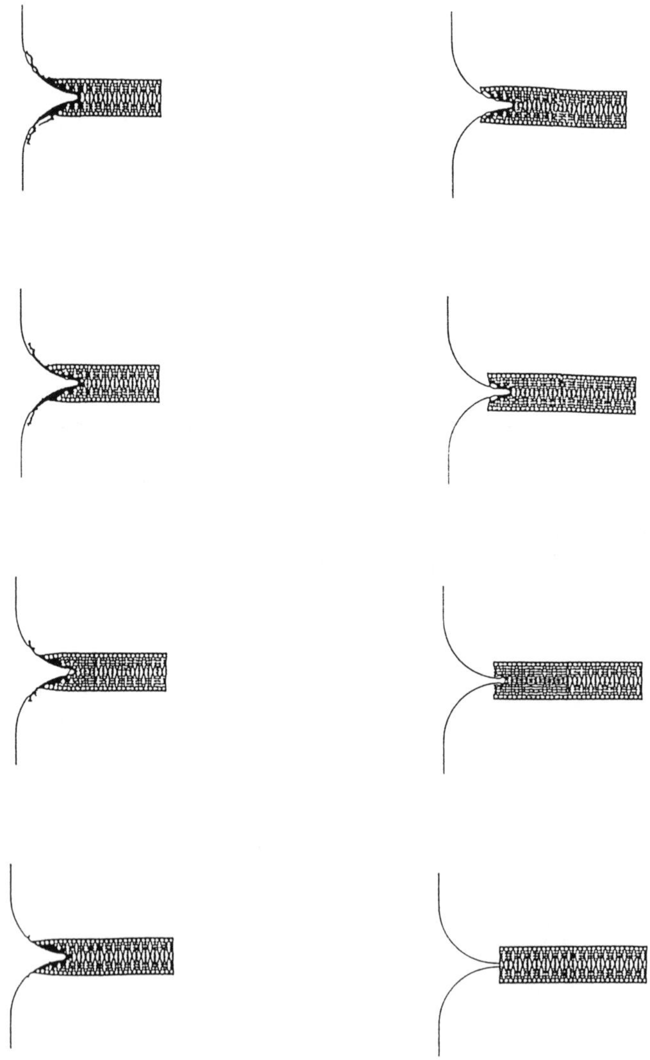

Figure 8.3–12: Skin contact with the edge for Test 3.

8.3–3 Car Front Energy Absorber (Bumper)

Background and Problem Description

In recent years there has been a large increase in the use of composite materials in the transportation industry in such a way that structural components usually made of metallic materials have been replaced by nonmetallic advanced materials. Nowadays crash simulation is becoming an efficient tool for the development of land transport vehicles. Crash simulations of all kinds of engineering structures like cars, trucks and railway cars made of steel, composites and other materials, can be achieved due to advances in numerical simulations and computer hardware and software. Moreover, the integration of numerical and experimental studies in the transportation industry has been made in order to achieve safety and efficiency in modern car design.

The material characterization is of special interest from the point of view of determining the global behavior. Strength, toughness, and mechanical properties of composite materials, in general, are so different from conventional models that new material models must be developed to define the behavior during the analysis of the problem. These models must also allow the evaluation of the development of damage and fracture initiation, propagation and instability, and enable the evaluation of strength and toughness criteria. Once the onset of failure or the damage of the material is determined, the next step consists of the identification of the mode of damage or failure and their quantification (*i.e.*, the classification).

A study of the energy absorption mechanism is presented in this section. The study is focused on the application of the results to a prototype of an absorbing element for the automotive industry. In order to save weight, the energy absorbers are replaced by parts made of advanced composite materials, which are structurally, as well as economically, competitive with the existing metal parts.

One of the applications of crash simulations in the automotive industry is energy absorbers, usually made of metallic materials. These elements, in case of crash of the vehicle, must absorb the crash energy in a given speed range first by reversible then by irreversible deformation of the absorbing device. Weight and manufacturing cost savings have been taken into account during the design of composite energy absorbers. Not only the change of material but also the physical effects (rolling energy, hydraulic energy, etc.) have been modified (deformation energy).

Designs of energy absorbers must be such that it is feasible to manufacture them in an easy way and if possible at low costs, and at the same time the function of the absorbing element must be simple. The main function of a crash absorber is to avoid damage to the rest of the structure. The damage in a structure is mostly due to the deformation energy (*i.e.*, energy required to deform the structure); thus it is necessary to build the structure with a mechanism that is able to absorb the deformation energy. Since the absorption energy depends on crash velocity, the speed ranges for reversible and irreversible deformation of the absorbing elements are a function of the design. The correlation between speed and energy absorption can be seen in Table 8.3–3.

Table 8.3–3: Speed range versus deformation energy for different markets.

Market	Deformation	
	Reversible speed range	Irreversible speed range
European	0 – 4 km/h	4 – 15 km/h
American	0 – 8 km/h	8 – 15 km/h

Some significant examples from the transportation industry can be chosen. In the automobile industry, more and more devices to avoid damage at low crash velocities, not only in the body work but also in the main structure, are being developed bearing in mind the safety of the occupants. In the aeronautical industry, a typical example that has been studied is a helicopter crash in case of power loss when the maximum vertical velocity is 15 m/s. There are two types of energy absorbers depending on the part of the structure in which we are interested. It may be feasible to put energy absorbers to reduce either the structural damage or the accelerations that the pilot should have to bear.

In both cases composite materials are being used in order to achieve light structures, but it is necessary to reduce manufacture and material costs. Therefore a competitive design must be performed. Numerical simulation of the impact and crash of structures made of composite materials has been carried out in the last several years and has demonstrated good agreement with experimental results. First, it is necessary to establish material models for composite strength and failure modes. In this case a biphase model has been used. Then, the structural configuration is calculated, taking into account the new material model. The results from this new configuration can be compared with experimental results. This simulation methodology is intended to predict the crash behavior of structures that could be subjected to impact. Once the behavior has been determined, it will allow us to optimize the design for crash damage and to make suitable changes, in order to prevent poor response.

In order to obtain such benefits some requirements should be satisfied: the structure must not suffer any kind of damage under a minimum load level, the structure must be able to reach a maximum load level before the failure initiation in the energy absorber will occur, in such a way that maximum load during the crash remains always below this level. Finally, a predetermined amount of energy, which depends on the maintenance of a high crash load and a large crash displacement, must be absorbed.

After a large number of numerical simulations of different specimens, validated with experimental testing, it could be demonstrated that composite materials are feasible alternatives to metals for energy absorbers in meeting the design objectives. In the following paragraphs the material selection, design considerations, and analysis aspects of energy absorbers are discussed.

Material

The shock absorber is composed of several types of materials, because of some requirements that have to be satisfied. Two important material systems are included in the design, the metallic and nonmetallic. The metallic system is used in the design of the support structure, and varieties of non metallic components are used as absorbing elements and friction devices.

Energy dissipating systems based on metallic structures have been developed in the past. The energy absorbing process in these systems is the formation of plastic zones in which the metal undergoes large strains in bending and stretching. The ability to undergo extensive slip without the presence of cracks is one of the characteristics of ductile metals such as mild steel and aluminum alloys. As the strain increases the slip process becomes more complicated and a higher stress is required to maintain the deformation process. The brittle behavior appears when the material is unable to slip extensively due to the formation of cracks. Mechanical response of a metal can be influenced by metal working processes such as alloying, mechanical deformation, and heat treatments during the slip process.

Some aspects of the deformation of metals are also applicable to plastics, although the details are quite different. Thermoplastics often undergo large plastic strains in tension before fracture, and thermoset plastics fail in tension after only two to six percent elongation. As is the case with metals, the strain rate sensitivity of plastics is related to the kinetics of motion of the molecules.

The part of the structure that is used as an energy absorber is a continuous fiber–reinforced composite tube. When it is subjected to crushing loads, the response is complex and depends on the interaction between the different mechanisms that control the crushing process. Three crushing modes can be identified for continuous fiber–reinforced composite tubes (see Figure 8.3–13):

1. Transverse shearing.
2. Lamina bending.
3. Local buckling.

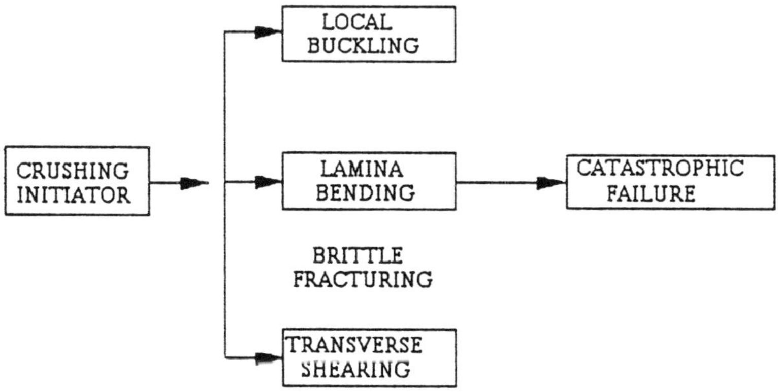

Figure 8.3–13: Crushing process of fiber–reinforced composite tubes.

Combination of fibers and matrices as well as the structure of the tube are important parameters to the progressive crushing of a composite tube. So not all composite tubes will progressively crush. Some mechanisms are able to cause catastrophic tube failures which avoid the progressive crushing. Catastrophic failures may occur, in a brittle fiber–reinforced composite tube where the length of the interlaminar cracks is less than a ply thickness or if internal cracks are quickly growing and propagating along the length of the tube.

Brittle fiber–reinforced composite materials exhibit exclusively transverse shearing and lamina bending crushing modes, but ductile fiber–reinforced composite materials do not show these modes. Local buckling–crushing mode can be exhibited by both ductile and brittle composites. A combination of the transverse shearing and lamina bending crushing modes can be exhibited by brittle fiber-reinforced composite tubes. The influence of material (resins and fibers), geometry, laminate stacking sequences, and velocity of the mechanism has been taken into account during the design stage.

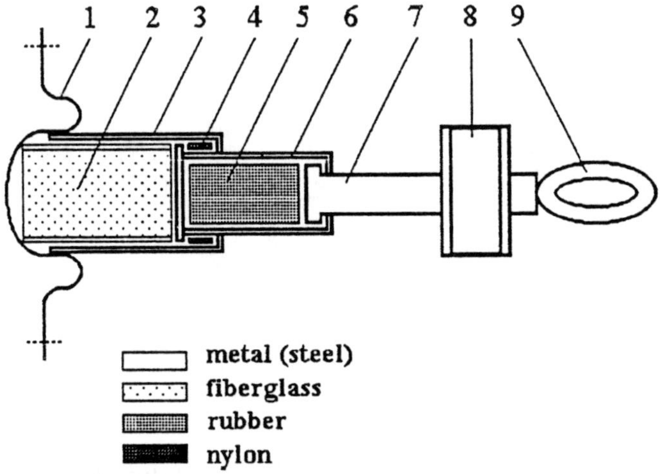

Figure 8.3–14: Prototype description and materials used.

Design Considerations

The design basically consists of nine parts which are made of four material types. It contains one internal composite tube, absorbing the energy during a crash of the car, and an external metal tube assuring the operation of the absorbing element and the stability of the structure during the tow of the car. Figure 8.3–14 shows the different parts and their material. As it can be seen, the prototype consists of metallic (steel) parts 1, 3, 6, 7, 8, and 9; composite parts (fiberglass) 2, nylon 4, and rubber 5. A tube is used in the design (number 2), which is made of fiberglass; this is the crash absorbing device during the irreversible deformation while a crash occurs. Different states of the shock absorber are shown in Figure 8.3–15.

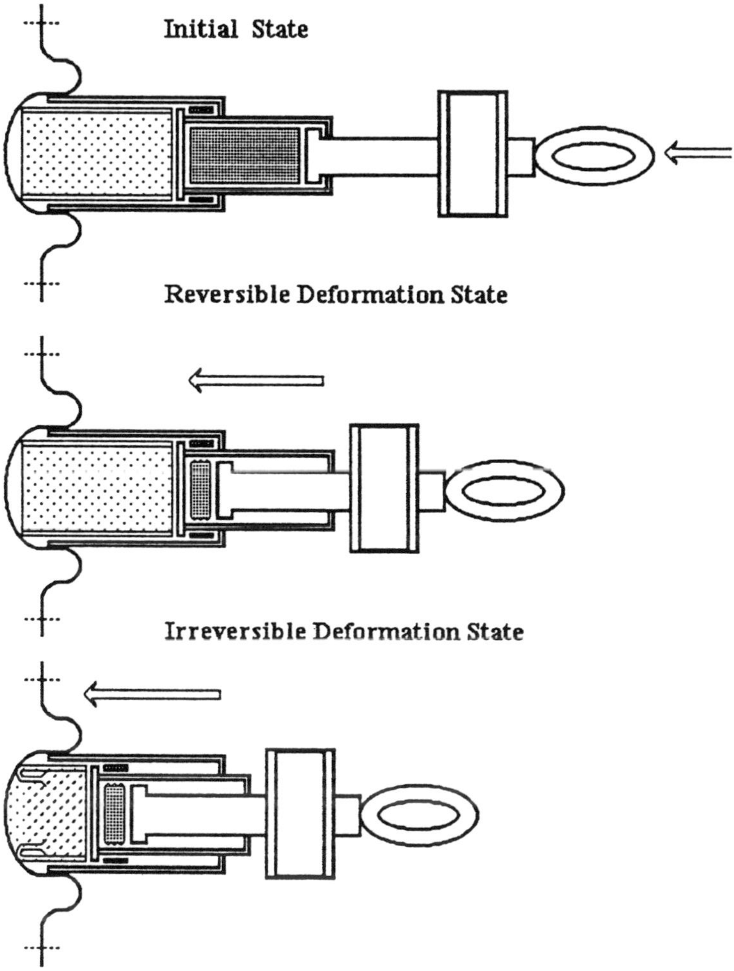

Figure 8.3–15: States during the shock absorber operation.

Reversible deformation: during a crash, parts 7, 8 and 9 move into the direction of part 6, compressing the rubber material (part 5) up to 25 mm, which is the maximum allowable reversible deformation of the absorber. If no more energy is introduced into the system, the rubber expands, and parts 7, 8, and 9 come back to their original position (see Figure 8.3–15).

Irreversible deformation: in the case of speeds from 4 to 15 km/h, part 7 touches part 6 and moves the structure consisting of parts 5, 6, 7, 8, and 9 against 1, which is fixed at the frame of the car, therefore destroying the composite cylinder (part 2) at the interface with part 1. The allowable irreversible deformation is 50 mm.

Interface: the interface between parts 1 and 2 has an important influence since geometrical relations between these parts are decisive for the initial failure behavior of the composite tube and therefore for assuring a stress-strain behavior similar to that of ductile materials.

Several designs of the interface have been studied in order to optimize the collapse of the cylinder with the objective of finding a geometrical relation that offers a stress-strain behavior as described before. The more important design variables that have been used can be summarized as follows:

Support Structure:

— geometry; maintaining some fixed dimensions

Absorbing element:

— tube length; ratio tube diameter/tube thickness; material configuration; material properties (fiber, resin, percentages); manufacturing process

Analysis

The numerical simulation of energy absorbers has been carried out by means of finite element computer codes. Models have been run in vectored and multitasked mainframe super computers, *e.g.*, CONVEX and HP. Several software packages have been used for numerical crash simulations, explicit finite elements programs with an interactive graphic meshing preprocessor and interactive graphic results postprocessing. Numerical simulations allow us to obtain new results that were hard to obtain, high priced, or impossible to reach, such as the detailed crash energy absorption histories of individual components. Descriptive pictures of particular and combined failure modes and the detailed history of the process are impossible to be observed by human eyes or high–speed cameras.

The analysis of the structure can be carried out by subdividing it according to the dominant behavior of the part that is being studied. The energy absorber can be divided basically into two structures, one of them, the absorbing element, has a dynamic dominant behavior whereas the support has mainly a static behavior. Then the analyses of the prototypes have been divided into two steps. One of these steps is related to the calculation of the support structure of the absorbing element, and the other is concerned with the dynamic analysis of the structural part that is subjected to crash.

The support has several requirements to be satisfied; one of these is to bear all the reactions transferred by the elements with dynamic behavior. If no crash occurs, the absorber may also serve as a towing device and must have a mechanism to support the bumper. However, the main task is to hold the components of the structure that have the role to absorb the energy by a reversible and an irreversible way and to assure the correct working of these parts. Therefore, two complementary analyses have been carried out, a static with large deformation analysis and a dynamic analysis.

The calculation of the crash absorber configuration has been performed using the substructure technique. Different substructures have been created to model the global structure, so several substructures were necessary to discretize the support structure, and the absorbing element, such as irreversible mechanism and reversible mechanism, have been independently treated. Then all of them are joined to complete global analysis of the structure.

Programs chosen for this task are DYNA3D and ABAQUS. Both contain truss, beam, shell, and solid elements that handle geometric and material nonlinearities. Dynamic processes are considered by the implementation of an explicit time integration algorithm, which is very fast and therefore suited for the calculation of time dependent problems as they occur in crash simulation. However, the analysis of composite materials is limited because the modelling of laminated plates is only considered by a homogenization procedure. Furthermore, a fracture criterion for dynamic processing is not available. These obstacles can be overcome since both codes are an open system and therefore manipulations of the source code by the user can be done easily.

Analysis of the support structure

Figure 8.3–16 shows the different substructures into which the prototype has been subdivided. A branch system is recommended in this case due to different behavior of each substructure and different material as well as different requirements that they should satisfy. Substructures 1 to 4 are metallic and 5 to 6 are made of composite materials. Substructures 1 to 4 and 6 are calculated as static problems with large deformations, and substructure 5 is calculated as a dynamic problem (crash analysis).

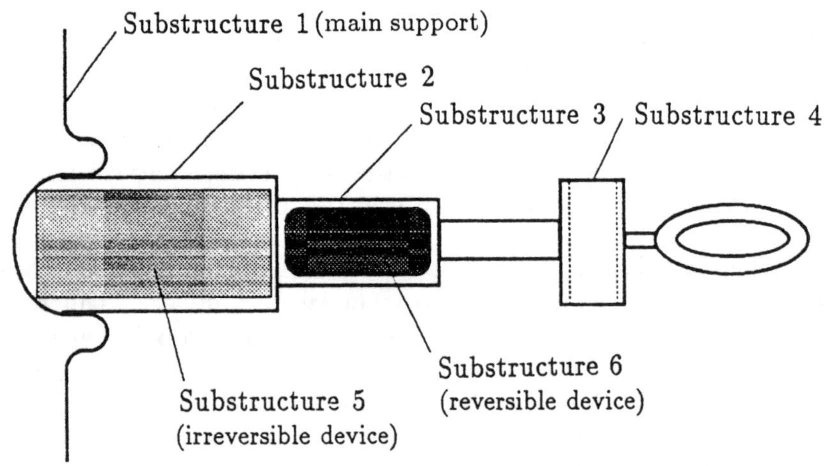

Figure 8.3–16: Schematic description of different substructures in which the shock absorber has been subdivided.

Support structure has been designed and calculated according to particular requirements. The main support is attached to the frame by means of bolted joints, and it possesses a mechanism to receive loads transmitted by the bumper and a rigid ring to serve as a towing device, if necessary.

A scheme of typical loads applied to support structure is shown in Figure 8.3–17. Values for axial load (La), vertical load (Lv) and horizontal load (Lh) are given according to speed range of the vehicle. Several load cases were analyzed depending on the use of the support structure and the way the design was optimized.

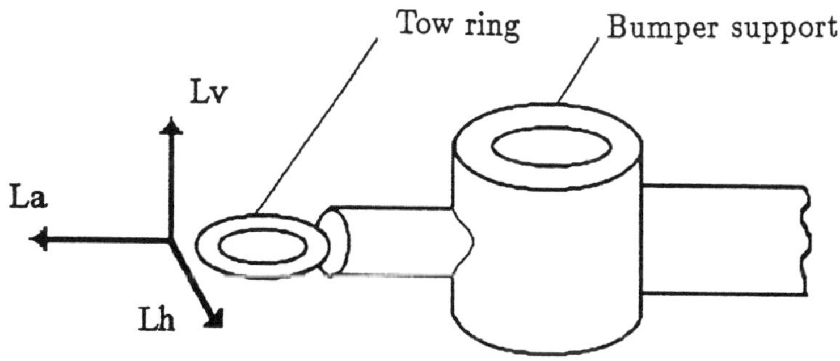

Figure 8.3–17: Applied loads to main structure.

Once the metallic support structure was calculated, stress concentrations were found near the bolted joints, welded parts, and junctions between different mechanical parts. The moving parts and the joints of the absorber are designed to withstand tension, bending, and shear loads while the car is towed. Large deformations were encountered when the structure was subjected to an extreme load, but these deformations were controlled by the design. A mechanism of this design controlled large deformations while the reversible device was working.

The calculation of this design under a load applied with an angle of attack of 70° was performed and could be considered as satisfying in terms of stiffness and strength. Figure 8.3–19 shows a stress and displacement mapping of the absorber under the load case described before. All the requirements were satisfied by the prototype, and bad responses were not found during the calculation. Furthermore, this design has also been checked in terms of a tensile load, and stress displacements were found to be completely acceptable.

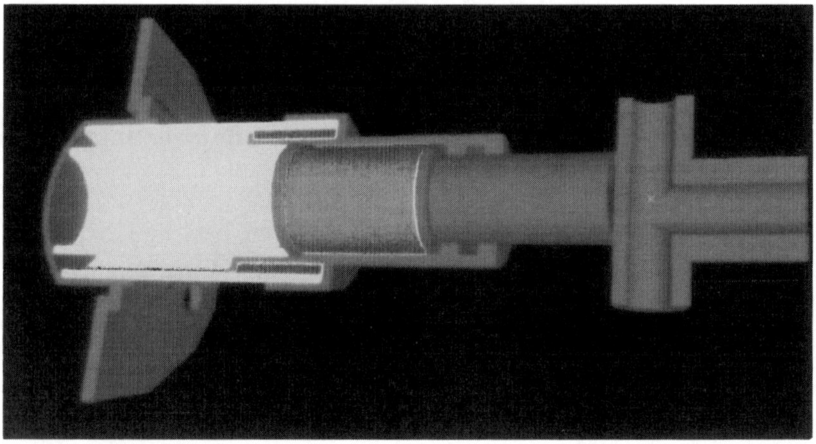

Figure 8.3–18: Model of the transverse section of the new energy absorber.

Figure 8.3–19: A classical shock absorber.

Analysis of the crash absorbing elements

The analysis of the crash absorbing elements has been carried out by means of the use of two finite element codes, DYNA3D and ABAQUS/Explicit. These packages are explicit, nonlinear, finite element codes for the transient dynamic response of three–dimensional solids and structures. They are recommended for problems where high rate dynamics or stress wave propagation effects are important. Both of them use a large number of relatively small time steps, and they use explicit solution methods, which are inexpensive, at each step, without forming and solving the large matrix equations, and do not require iterations at each step. Large computing time savings can be achieved due to the above mentioned goals. An implicit code often needs a lot of memory due to the assembly of a large stiffness matrix at each time step. These finite element codes are based on a finite element discretization of the three spatial dimensions and a finite difference discretization in time. The method used to integrate the equations of motion in time is the explicit central difference method. Automatically, the maximum time step size at each step of the solution is calculated by the codes, and the time step is adjusted in order to minimize the number of time steps used in a solution.

Experiment

A large number of experiments were conducted to control the required parameters such as maintained load, testing velocity, etc. In Figure 8.3–20 a view of the experimental testing can be seen. Several types of tests were conducted, depending on the parameters to be controlled. An important parameter was the design and the calculation of crash initiators in order to obtain the desired force-displacement curve.

Energy absorber device crash tests

The typical load–displacement curve for cylindrical composite tubes without any failure initiator shows an initial high load and a rapid loss of load carrying ability. Inclusion of a load limiting failure initiator at one of the ends results in a loss or a reduction of the initial peak load but enhances the post failure performance. Thus, in order to obtain the best maintained load and a progressive crash behavior, the inclusion of triggering devices is needed.

Figure 8.3–20: A picture of a crash test.

Three different types of crash initiators have been used in this study, as shown in Figure 8.3–21. Type A is a chamfer that has been tested with three different angles (30, 45 and 60 degrees) in two different versions, inside and outside chamfer. Load–displacement curves are shown in Figures 8.3–22 through 8.3–24. The only influence of the chamfer–angle is how quickly the stable crash load will be reached. The amount of this maintained load seems to be independent of the angle. The results obtained with the inside and outside chamfer are very similar.

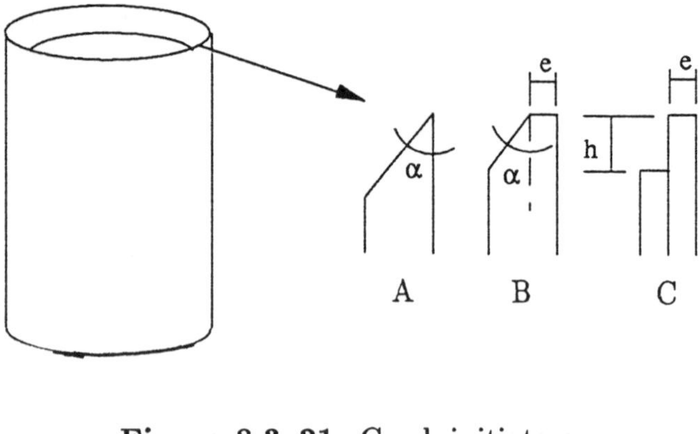

Figure 8.3–21: Crash initiators.

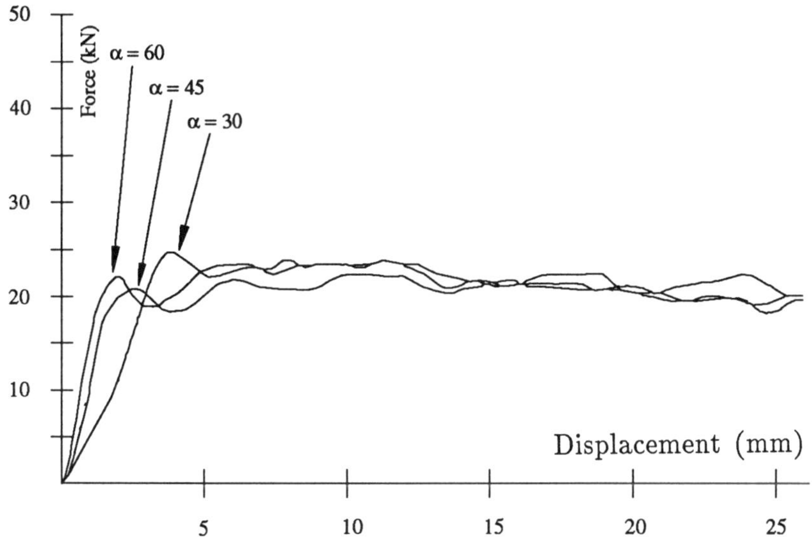

Figure 8.3–22: Force–displacement curves for tubes with triggering device type A. Influence of the angle α.

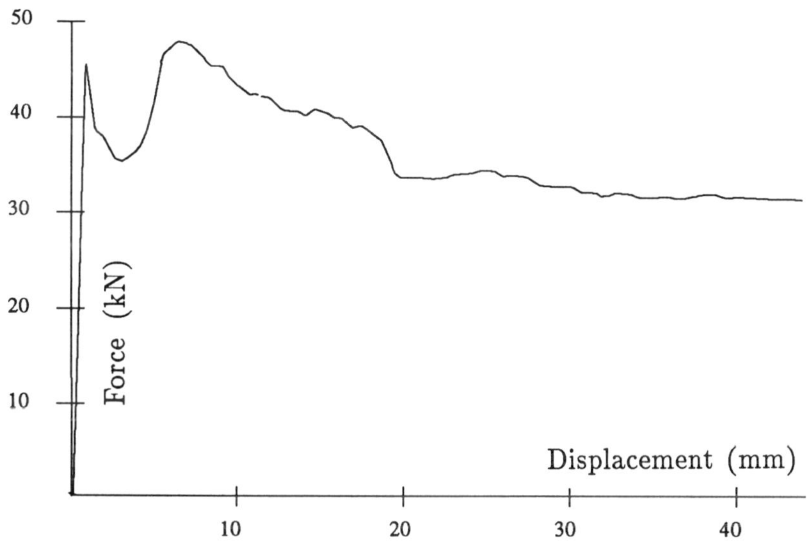

Figure 8.3–23: Force–displacement curves for tubes with triggering device type B.

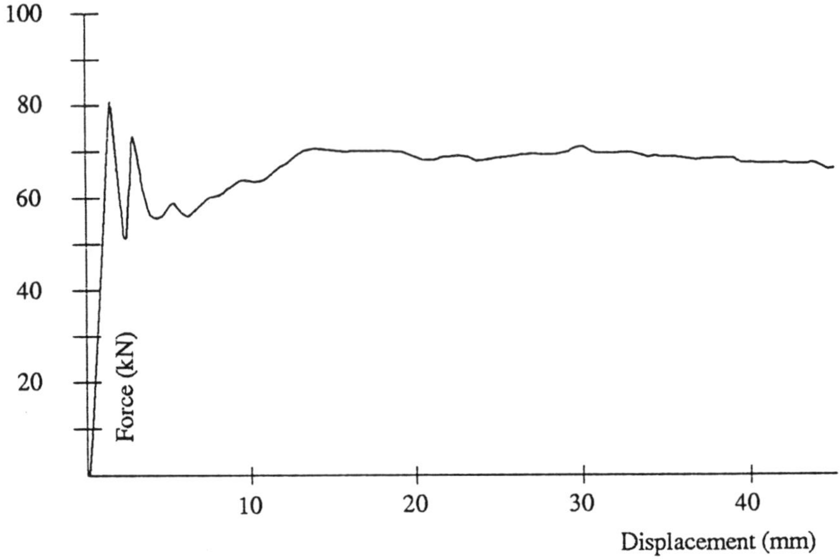

Figure 8.3–24: Force–displacement curves for tubes with triggering device type C.

Type B is a chamfer in the half inside or outside thickness. The angles used are 30, 45 and 60. The objective of this kind of triggering device initiator is to reach the maximum reversible load without damage. Predicting the initial peak load is not simple; it depends on the lay–up configuration in the nonchamfered section. The initial part of the crash testing could be unsteady with a progressive reduction in the maintained load.

Type C is a group of two tubes with different lengths. The influence of the difference between the two tube lengths (L2–L1) has been observed. The force–displacement curve for this case has two load peaks. The influence of L2–L1 is represented by the force at the second peak. It has been found that L2–L1=2 mm leads to the best behavior. The amount of the two peaks is nearly equal and only slightly superior to the crash maintained load.

Results

An example of the results obtained during the analysis process can be seen in Figure 8.3–25. Correlation has been found between theoretical (numerical simulations) and experimental (testing) results. This graph shows us a good agreement between the two methods that have been used to analyze the behavior of the crash absorbing parts.

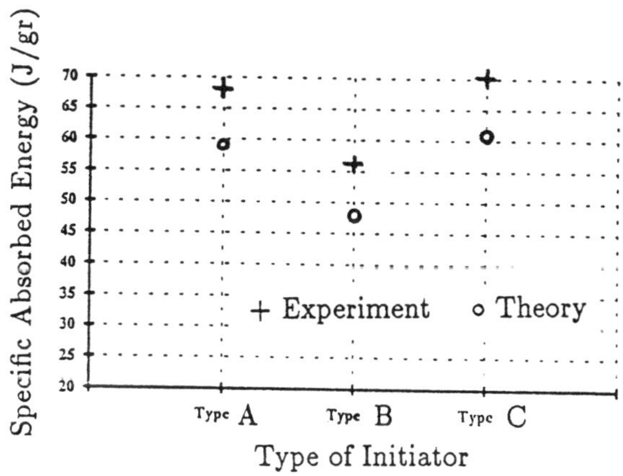

Figure 8.3–25: Numerical–experimental correlation.

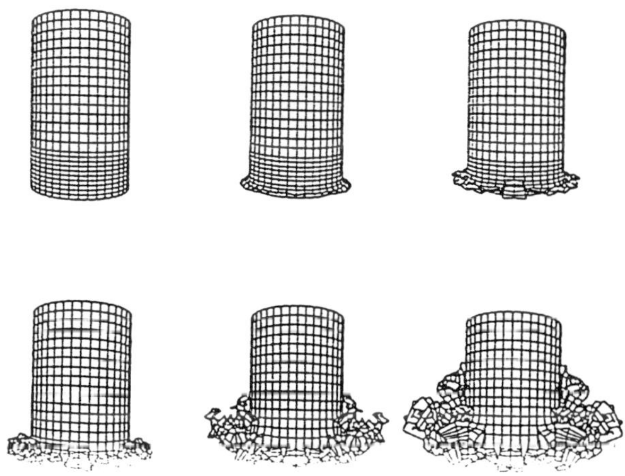

Figure 8.3–26: Numerical simulation of crushing of a tube.

The sequence shown in Figure 8.3–26 was obtained during the numerical simulations with the finite element computer codes mentioned earlier. Six deformation stages have been extracted from the numerical simulation of the irreversible part (a composite tube) of the shock absorber. Numerical simulations can be compared with experimental results as shown in Figure 8.3–27. The same testing conditions were used during the experimental testing, and the results proved the power and the utility of these numerical tools.

Figure 8.3–27: Experimental crushing of a tube.

Another result that can be brought up is that composite materials are feasible to be included in mechanisms similar to those mentioned in this study. So, a desirable behavior can be obtained with the use of composite material, as can be seen in Figure 8.3–28. The load and displacement requirements are completely satisfied by the presented design.

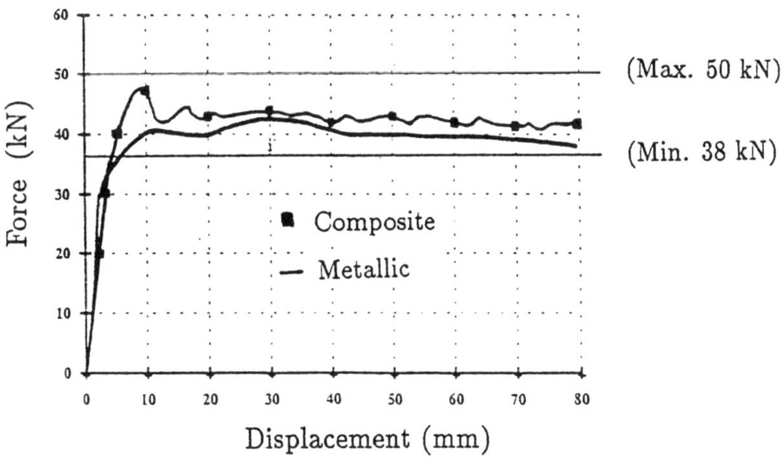

Figure 8.3–28: Metallic and non–metallic comparison.

Conclusions

The method of calculation that has been developed gives a good agreement between theoretical and experimental results. Theoretical simulations help us to reduce the amount of experimental testing. Composite materials are feasible to be used as a choice in crash absorbing devices because of their stable crash. A competitive design has been obtained due to their good behavior and easy manufacturing, so weight savings and low costs are two of the most prominent characteristics.

REFERENCES FOR ADDITIONAL READING

1. Reddy, J. N. and Rasmussen, M. L., *Advanced Engineering Analysis*, John Wiley & Sons, New York, 1982; reprinted by Krieger, Melbourne, FL, 1990.

2. Reddy, J. N., *Energy and Variational Methods in Applied Mechanics*, John Wiley & Sons, New York, 1984.

3. Malvern, L. E., *Introduction to the Mechanics of a Continuous Medium*, Prentice–Hall, Englewood Cliffs, NJ, 1969.

4. Reddy, J. N., *An Introduction to the Finite Element Method*, Second Edition, McGraw–Hill, New York, 1993.

5. Zienkiewicz, O. C. and Taylor, R. L., *The Finite Element Method,* McGraw–Hill, New York, 1989.

6. Hull, D., "Energy Absorbing Composite Structures, Scientific and Technical Review," University of Wales, No. 3, pp. 23–30, 1988.

7. Whirley, R. G. and Hallquist J. O., "DYNA3D: Non–Linear, Explicit, Three–Dimensional Finite Element Code for Solid and Structural Mechanics–User Manual," Lawrence Livermore National Laboratory, Livermore, CA, 1991.

8. ABAQUS, Version 4.9, Hibbitt, Karlsson and Sorensen, Inc., Pawtucket, RI.

9. Thorton P. H. and Jeryan R. A., "Crash Energy Management in Composite Automotive Structures," *Int. Journal Impact Engineering,* **7**(2), 167–180, 1988.

10. Farley, G. L. and Jones, R. M., "Crushing Characteristics of Continuous Fiber–Reinforced Composite Tubes," *Journal of Composite Materials,* **26**(1), 1992.

11. Jones, R. M., *Mechanics of Composite Materials,* Scripta Book Co., Washington, DC, 1975.

12. Zienkiewicz, O. C., *El método de los elementos finitos,* Ed. Reverté, 1980.

13. Tsai S. W., *Theory of Composite Design,* Think Composites, Dayton, OH, 1992.

14. Liu Xia–Shi, "Finite Element Analysis for Large Composite Structure," *Proceedings of Fourth International Conference on Composite Structures (ICCS-4),* I. H. Marshall (ed.), Applied Science Publishers, New York, NY, 1987.

15. Thornton, P. H., "Energy Absorption in Composite Structures," *J. Compos. Mater.,* **13,** 247–262, 1979.

16. Kirsch, P. A. and Jahnle, H. A., "Energy Absorption of Glass–Polyester Structures," SAE Paper 810233, 1981.

17. Thornton, P. H. and Edwards, P. J., "Energy Absorption in Composite Tubes," *J. Compos. Mater.,* **16,** 521–545, 1982.

18. Hull, D., "Energy Absorption of Composite Materials under Crash Conditions," *Progress in Science and Engineering of Composites,* T. Hayashi, K. Kawata and S. Umekawa (eds.), ICCM–IV, **1**, Tokyo, 861-870, 1982.

19. Farley, G. L., "Energy Absorption of Composite Materials," *J. Compos. Mater.,* **17**, 267–279, 1983.

20. Kindervater, C. M., "Energy Absorbing Qualities of Fiber Reinforced Plastic Tubes," National Specialists Meeting of Composite Structures of the American Helicopter Society, 1983.

21. Hull, D., "Axial Crushing of Fiber Reinforced Composite Tubes," *Structural Crashworthiness,* N. Jones and T. Wierzbicki (eds.), Butterworth & Co. Ltd. (UK), London, 118–135, 1983.

22. Snowdon, P. and Hull, D., "Energy Absorption of SMC Under Crash Conditions," *Fiber Reinforced Composites Conference,* Plastics and Rubber Institute, pp. 5.1–5.10, 1984.

23. Berry, J. and Hull, D., "Effect of Speed on Progressive Crushing of Epoxy–Glass Cloth Tubes," *Mechanical Properties of Materials at High Rates of Strain,* J. Harding (ed.,) Inst. Phys. Conf. Series No. 70, New York, 463–470, 1984.

24. Keal, R., "Post Failure Energy Absorbing Mechanisms of Filament Wound Composite Tubes," Ph.D. Thesis, University of Liverpool, Liverpool, UK, 1984.

25. Berry, J. P., "Energy Absorption and Failure Mechanisms of Axially Crushed GRP Tubes," Ph.D. Thesis, University of Liverpool, Liverpool, UK, 1984.

26. Thornton, P. H., "Effect of Trigger Geometry on Energy Absorption in Composite Tubes," Proc. of the Fifth Int. Conf. Composite Materials (ICCM–V), W.C. Harrigan, S. Strufe, and A. K. Dhingra (eds.), Metallurgical Society, Warrendale, PA, pp. 1183–1199, 1985.

27. Thornton, P. H., "The Crush Behavior of Glass Fiber Reinforced Plastic Sections," *Composites Science Technology,* **27**, 199–224, 1986.

28. Farley, G. L., "Effect of Fiber and Matrix Maximum Strain on the Energy Absorption of Composite Materials," *J. Comp. Mat.,* **20**, 322–334, 1986.

29. Bravo, P., Larrode, E., and Miravete, A. "Crash Absorbing Elements for Automotive Applications," *Proceedings ICCM/9*, Vol. 5, pp. 374–385, A. Miravete (ed.), Madrid, 1993.

30. Bravo, P., Larrode, E., Miravete, A., and Tamparillas, S. "Crash Absorbing Elements for Automotive Applications," *International Journal of Composite Structures*, to appear.

31. Bravo, P. "Anáslisi y diseño de absorbedores de energía con materiales compuestos y su aplicación a los medios de transporte," Ph.D. Thesis, University of Zaragoza, Spain, 1994.

32. Larrode, E., Miravete, A., Bravo, P., and Baselga, S., "Design and Analysis of Crash Absorber Made of Advanced Materials: Automotive Applications," *Proceedings of ISATA-27,* Aachen, Germany, 1994, pp. 199–206.

Chapter Nine

ANALYSIS OF LARGE STRUCTURES

9.1 INTRODUCTION

The finite element method has emerged as the most powerful numerical technique ever devised to analyze practical engineering structures. The method is endowed with a basic feature, *divide the whole into parts that are simple to work with* [1–5], which makes the method applicable to a variety of physical phenomena, and enables the development of *general purpose* computer programs that can be used for *any* geometry, physically meaningful boundary conditions, and loading. However, many of the modeling considerations, such as the selection of the element and mesh used, and problem formulative efforts depend, among other things, on a good understanding of the physics of the problem being analyzed, the required accuracy, the available computational resources, and the intended use of results. Ultimately, the finite element results are only as accurate as one's physical and numerical representation of the problem. A realistic representation of the actual problem will yield a very good correlation between the analysis results and experimental data.

For some systems, a crude approximation may be sufficient to give a good idea about their response. There are other systems that must be analyzed with care and with a fine mesh. Since computational resources are limited, the mesh refinement is limited by the computer storage and time of computation. In such cases, the response of the system may be determined more accurately by using refined meshes only in those critical regions where the solution gradients are large. Such information is qualitatively known to an experienced engineer who is knowledgeable about the problem. The global–local modeling and substructuring techniques may be used to solve some problems that, due to their complexity or size, would require a large number of elements to obtain a solution.

Certain composite material structures, because of their complexity, need to be analyzed in detail; such structures may be analyzed using the substructuring technique. This technique allows us to isolate zones that require accurate information (or critical zones) and simplify the problem by reducing solution times and computer storage space. Some examples of modeling issues associated with the analysis of practical problems are outlined below.

- *Reduction of problem size:* when the size of the problem is very large, it is necessary to reduce the number of variables because of solution time and storage requirements (*e.g.*, layer by layer analysis, repetitive analysis).

- *Detailed studies:* A specific part of the model needs to be analyzed in detail and a refined meshing would increase the number of variables (*e.g.*, joints, holes, free edges, crack growth, delaminations).

- *Multiple solutions:* Due to long time solutions, linear and nonlinear parts must be isolated in order to reduce computing time (*e.g.*, local buckling, interface contact analysis, dynamic analysis, etc).

- *Analysis by parts:* When a structure is composed of several distinct parts, they could be designed/analyzed separately at different stages or by different groups, or some of these parts may already have a fixed design and others do not.

9.2 THE SUBSTRUCTURING TECHNIQUE

Substructuring consists of the division of a single structure into several substructures. One of the goals of the substructuring technique is to obtain a complete and detailed information about the response of a large structure under different loads by studying individual substructures. If the structure requires a large number of finite elements to represent its response accurately, then a substructuring technique can be used to analyze each substructure and to avoid taxing the available computer capacity.

The general procedure of the substructure technique can be summarized as follows. First, the part that is selected to be repeated along the complete model can be modeled as usual; the model is saved as a set of master degrees of freedom. Second, the substructure matrices are calculated and written to a file. Once the substructure has been formulated, it may be used by itself in the same way as other standard finite elements. Several advantages of the substructuring technique are summarized below.

- Substructuring can be used as an efficient optimization tool. When the design needs to preserve certain parts without modification, these can be modeled as substructures and therefore isolated from the rest of the structure. During the analysis they remain unchanged if any process of redesigning or mesh refinement is applied to the rest of the model.

- The process of modeling can be accomplished independently by several analysts, and the individual parts can be put together to construct the complete model.

- Other important advantages are the savings in modeling and solution times for a structure having repeated or symmetric parts. One of these parts can be formed as a substructure, which is then repeated to form the entire structure.

- Time saving during the calculation can be achieved not only by means of using repeated substructures but also by the ability to separate the stress and displacement calculations, which can be processed sequentially as a postprocessing step.

- Once the substructure has been modeled and formulated, different types of analyses can be performed on the same substructure without having to form or re-tridiagonalize the stiffness matrix.

- It is possible to separate the linear and nonlinear parts of the structure, so that linear parts of the model are not subjected to an iterative solution.

The substructuring technique is based on matrix condensation; the assembled system of equations (hence, the stiffness matrix and the mass matrix) is reduced to a system of equations composed only by the master degrees of freedom. The set of master degrees of freedom must be selected carefully because they define the behavior of the structure being analyzed. A brief description of the condensation procedure and recovery is presented next.

Consider part B of the structure shown in Figure 9.2–1, which is identified as a substructure. The equilibrium equations of the substructure can be written in matrix form as

$$[K]\{\phi\} = \{F\} \tag{9.2 – 1}$$

where $[K]$ is the (symmetric) stiffness matrix, $\{\phi\}$ is vector of nodal displacements, and $\{F\}$ is the external force vector.

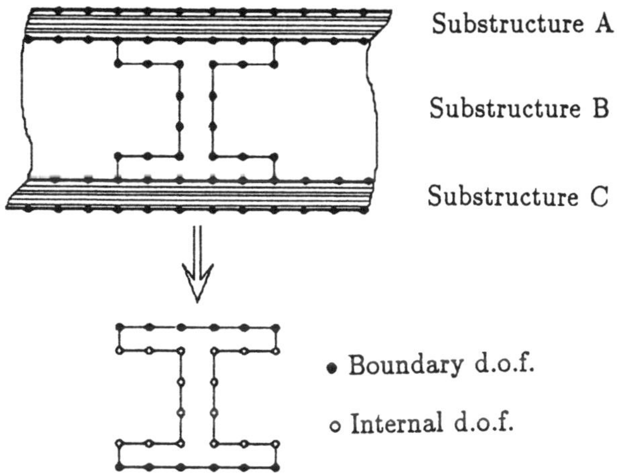

Figure 9.2–1: Example of the condensation process.

Substructure B is composed of internal nodes i and boundary nodes b. The boundary nodes are those that connect the substructure B to the rest of the structure. Thus, Eq. (1) can be rewritten in partitioned form as

$$\begin{bmatrix} [K_{ii}] & [K_{ib}] \\ [K_{bi}] & [K_{bb}] \end{bmatrix} \cdot \begin{Bmatrix} \{\phi_i\} \\ \{\phi_b\} \end{Bmatrix} = \begin{Bmatrix} \{F_i\} \\ \{F_b\} \end{Bmatrix} \tag{9.2 – 2}$$

or

$$[K_{ii}]\{\phi_i\} + [K_{ib}]\{\phi_b\} = \{F_i\} \qquad (9.2-3)$$

$$[K_{bi}]\{\phi_i\} + [K_{bb}]\{\phi_b\} = \{F_b\} \qquad (9.2-4)$$

whereϕ_b are the boundary degrees of freedom to be retained and ϕ_i are the degrees of freedom to be eliminated. Solving Eq. (3) for ϕ_i in terms of ϕ_b, we obtain

$$\{\phi_i\} = [K_{ii}]^{-1}\{F_i\} - [K_{ii}]^{-1}[K_{ib}]\{\phi_b\} \qquad (9.2-5)$$

and substituting the result into Eq. (4), we obtain the condensed equations

$$[\tilde{K}_b]\{\phi_b\} = \{\tilde{F}_b\} \qquad (9.2-6)$$

where

$$[\tilde{K}_b] = [K_{bb}] - [K_{bi}][K_{ii}]^{-1}[K_{ib}] \qquad (9.2-7a)$$

$$\{\tilde{F}_b\} = \{F_b\} - [K_{bi}][K_{ii}]^{-1}\{F_i\} \qquad (9.2-7b)$$

With this expression we are able to define a transformation that relates the load vector $\{\tilde{F}_b\}$ (the condensed load vector) with the displacements of the boundary nodes $\{\phi_b\}$, and $[\tilde{K}_b]$ denotes the condensed stiffness matrix.

In the same way, similar expressions can be found for the substructures A and C. Successively it is possible to establish the equilibrium equations in matrix form at the common boundaries of the three substructures. The assembling operation of the substructures can be performed, considering them as standard finite elements and calculating the displacements of the degrees of freedom of the common boundary nodes of the substructures. This set of assembled substructures generates the main structure, which is defined by the degrees of freedom belonging to the common boundary nodes of substructures plus those that do not belong to the condensed part of the structure if they exist. Then the equilibrium equations can be written as

$$\{\tilde{F}_b\}_A\{\tilde{F}_b\}_B\{\tilde{F}_b\}_C = \{F_b\}_T \tag{9.2-8}$$

where F_{bT} is the external load vector acting in the common boundaries of the three substructures. Once the total structure has been calculated, the displacements of the internal nodes of the different substructures can be recovered, and, following an inverse method, the internal displacements can be obtained. In view of Eq. (2), we can express the equilibrium equation for each substructure affected by others as

$$\begin{bmatrix} [K_{ii}] & [K_{ib}] \\ [K_{bi}] & [K_{bb}] \end{bmatrix} \begin{Bmatrix} \{\Phi_i\} \\ \{\Phi_b\}_T \end{Bmatrix} = \begin{Bmatrix} \{F_i\} \\ \{F_b\} \end{Bmatrix} \tag{9.2-9}$$

where $\{F_b\}_T$ can be obtained by adding external loads on the boundary nodes of the substructure to the reactions transferred by the other substructures at these nodes.

Solving Eq. (9) for Φ_i, we obtain

$$\{\Phi_i\} = [K_{ii}]^{-1}\{F_i\} - [K_{ii}]^{-1}[K_{ib}]\{\Phi_b\} \tag{9.2-10}$$

Equation (5) can be used to obtain $[K_{ii}]^{-1}\{F_i\}$

$$[K_{ii}]^{-1}\{F_i\} = \{\phi_i\} + [K_{ii}]^{-1}[K_{ib}]\{\phi_b\} \tag{9.2-11}$$

Next we need to know the boundary displacement vector $\{\phi_b\}$ for each substructure. By applying a unit load to the boundary degrees of freedom, $\{\phi_b\}$ can be calculated. Using $\{F_b\} = \{1\}$ and $\{F_i\} = \{0\}$ in Eq. (2), we obtain

$$\begin{bmatrix} [K_{ii}] & [K_{ib}] \\ [K_{bi}] & [K_{bb}] \end{bmatrix} \cdot \begin{Bmatrix} \{\phi_i'\} \\ \{\phi_b'\} \end{Bmatrix} = \begin{Bmatrix} \{0\} \\ \{1\} \end{Bmatrix} \tag{9.2-12}$$

Solving this system, we obtain

$$\{\phi_i'\} = -[K_{ii}]^{-1}[K_{ib}]\{\phi_b'\} \tag{9.2-13}$$

$$\left([K_{bb}] - [K_{bi}][K_{ii}]^{-1}[K_{ib}]\right)\{\phi_b\} = \{1\} \tag{9.2-14}$$

Equation (14) yields [*cf*. Eq. (7a)]

$$[\tilde{K}_b] = \{\phi'_b\}^{-1} \tag{9.2 – 15}$$

As in Eq. (6), we obtain

$$\{\tilde{F}_b\} = \{\phi'_b\}^{-1}\{\phi_b\}\ , \quad \{\phi_b\} = \{\phi'_b\}\{\tilde{F}_b\} \tag{9.2 – 16}$$

Substituting in Eq. (11), we obtain

$$[K_{ii}^{-1}]\{F_i\} = \{\phi_i\} + [K_{ii}^{-1}][K_{ib}]\{\phi'_b\}\{\tilde{F}_b\} \tag{9.2 – 17}$$

and using Eq. (13), we arrive at the result

$$[K_{ii}]^{-1}\{F_i\} = \{\phi_i\} - \{\phi'_i\}\{\tilde{F}_b\} \tag{9.2 – 18}$$

Finally, returning to Eq. (10), we have

$$\{\Phi_i\} = \{\phi_i\} - \{\phi'_i\}\{\tilde{F}_b\} - [K_{ii}]^{-1}[K_{ib}]\{\Phi_b\} \tag{9.2 – 19}$$

There is an important benefit of the substructuring technique when dealing with laminated composite materials. The sublaminate that repeats in a laminated structure can be treated as a substructure, so the laminate can be constructed in an easy way. The number of plies, orientation of each lamina, as well as the material property of each ply are then a part of the substructure (see Figure 9.2–2). A library of sublaminates (substructures) can be created, and, to simplify the selection of the sublaminates during the optimization phase, they can be ordered and classified. An effective method to realize computing and modeling time savings is to use libraries of substructures, which correspond to certain sublaminates.

The substructure can be modeled using the standard finite element procedures. Each lamina can be modeled with finite elements. Thus, a first analysis determines nodal forces at nodes on boundaries. These nodes are the master nodes that join the substructure with others. Depending on the degrees of freedom selected and the type of analysis, the approach may vary from a coarse approximation (selecting a few degrees of freedom) to a refined one (selecting all nodes with all their degrees of freedom). An example of the standard finite element mesh discretization and its condensed model are shown in Figure 9.2–3. The substructure contains only boundary nodes.

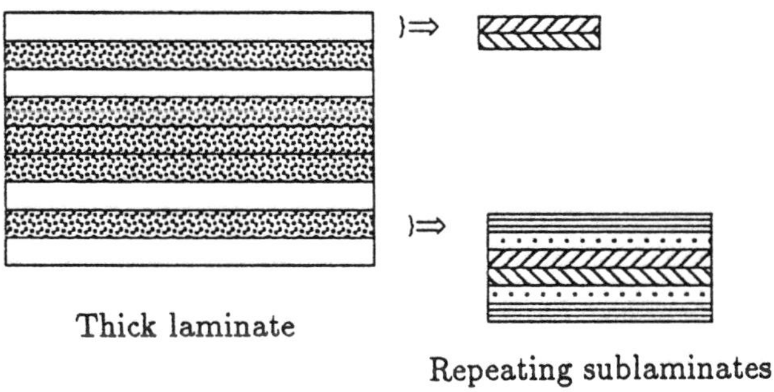

Figure 9.2–2: Substructuring of laminates.

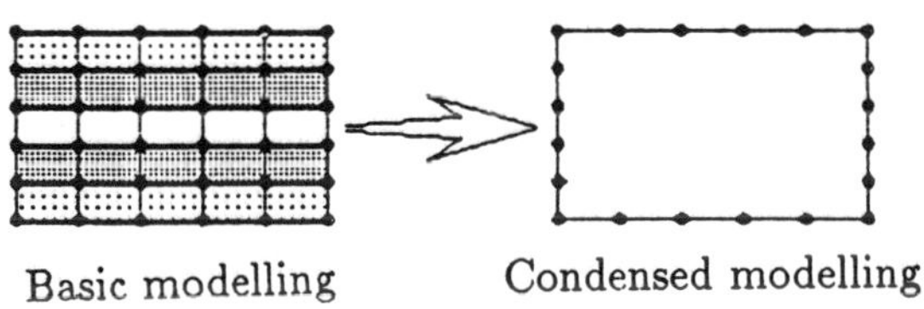

Figure 9.2–3: Mesh generation in laminates.

Once the sublaminate has been modelled as a substructure, nodal displacements on the interface boundaries are calculated by solving the equations of the system. After the boundary nodal displacements are obtained, the interface degrees of freedom of the sublaminate (substructure) are recovered, and they are used to analyze the sublaminate with both sets of degrees of freedom and known loads at internal degrees of freedom.

Depending on the type of laminate stacking, we can construct different groups in order to have an optimum number of plies in each substructure. For example, six plies per substructure were found to be the optimum during the analysis of a laminated plate (Figures 9.2–4 and 9.2–5).

Optimum Laminates and Optimum Grouping

$$(45/-45)_{18} = (45/-45/45/-45/45/-45)_6$$
$$(0/90/45/-45)_9 = (a/b/a/b/a/b)$$
$$(0_2/90_2/45/-45)_6 = (0/0/90/90/45/-45)_6$$

$$a = (0/90/45/-45/0/90)\ ,\ b = (45/-45/0/90/45/-45)$$

Figure 9.2–4: Example of laminate grouping to optimize computing time.

A drawback of the substructuring technique is that the stiffness matrix of a substructure is a full matrix, whose size may be large if the substructure has a large number of condensed degrees of freedom. This may represent an increase of the wavefront of the model leading to large computing times for solution of the equations. Therefore, it is necessary to find an appropriate number of boundary nodes and their degrees of freedom carefully; several small substructures might be a better choice than a single large substructure. An optimum number of substructures which require the minimum computing time could be found, as shown in Figure 9.2–6.

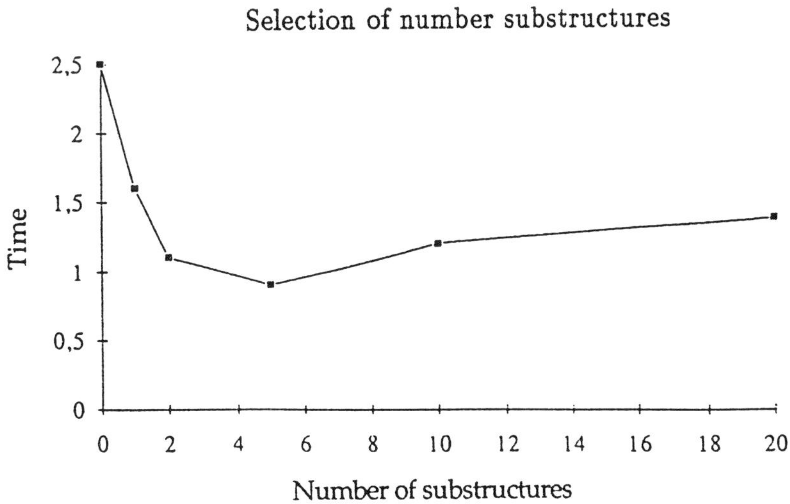

Figure 9.2–5: Optimum number of plies per substructure.

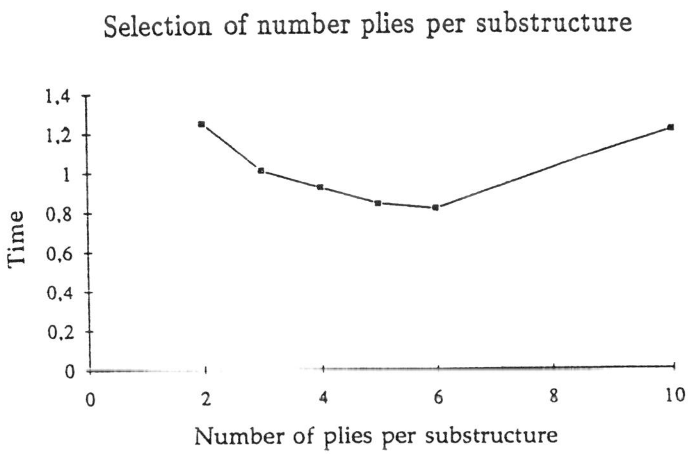

Figure 9.2–6: Optimum number of substructures.

9.3 ANALYSIS CONSIDERATIONS

9.3–1 Analysis Aspects

The design of structures made of composite materials is carried out by taking into account the large number of factors that make the process more complicated because of a large number of variables required. The stacking sequence, types of joints (*e.g.*, adhesive, bolted, etc.), and other factors increase the size of the problem greatly. Therefore, a powerful analysis tool to manage all the variables of the design is required. The finite element method is usually selected to carry out this task. Moreover, workstations rather than mainframe computers provide the possibility of solving the problem in a short period of time. Computational advantages can be realized if the substructuring technique is used to analyze large composite structures. Simple models for each substructure can be identified prior to the definitive modeling in order to know their behavior and to assess CPU time and computer storage requirements. An initial simple analysis of the whole model gives us an idea of the behavior of the model, and it indicates the critical areas of the design.

Once the initial analysis is completed, the next step consists of the mesh generation of substructures. Components of the main structure can be analyzed separately and optimized according to the general requirements of the design. Later, all the parts are brought together to form the full model. Important aspects during the modeling are the location of the nodes and elements connectivity among different substructures. A careful selection of the nodes that join the substructures to each other has to be made in order to avoid discontinuities. A continuous communication between people who model different substructures must be established so that information about the position of boundary nodes is taken at each stage during modeling. The number of degrees of freedom of each substructure is decisive for the approach to be adopted.

Once the whole model is constructed a complete analysis can be carried out to obtain the results. These results are then returned to the different substructures and suitable modifications are made on each substructure. An iterative process of optimization is then established until the main objective is achieved. After the iterative process ends, the results provided by the analysis are compared with experimental data, and an optimum design of the structure is obtained. Information about the behavior of the components of a structure is studied separately. The number of nodes as well as their location in the discrete model of the main structure can also be examined. It should be noted that every change in a part of the structure affects the other parts, therefore modifications must be done carefully; this is why an analysis of the entire model in each step during the process is very helpful.

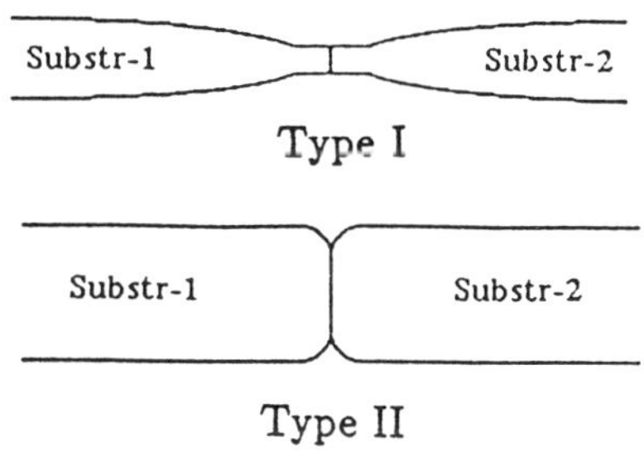

Figure 9.3–1: Types of boundary connections.

When analyzing some structures there are two possibilities to determine the location of boundary nodes (see Figure 9.3–1): fixed and nonfixed boundary nodes, which are influenced by the way the substructures are connected. That is,

- *Fixed boundary* (Type I) nodes are those for which there is no doubt in selecting the appropriate number of boundary degrees of freedom because the separation strip among two substructures is very short, and the number of degrees of freedom does not increase the number of variables of the problem.

- *Nonfixed boundary* (Type II) corresponds to the case where a selection of main degrees of freedom is feasible in order to decrease the number of variables.

9.3–2 Substructuring Work Scheme and Procedure

A detailed workplan concerning design, modeling, optimization, computation, and verification must be carried out. As a result, an efficient procedure to analyze large structures made of composite materials may be developed, conclusions about the behavior of the structure and the composite materials used may be obtained, and advantages of the optimization resulting from the process may be assessed.

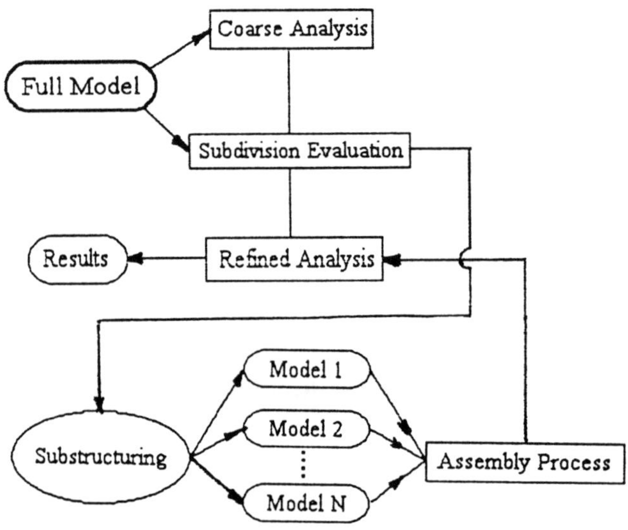

Figure 9.3–2: Flow Diagram.

During the preparation of tasks prior to the start of the analysis of the structure, several preliminary schemes are performed to carry out each task in an effective manner. During this stage modifications and changes to the plan may be made. A flow chart of the general procedure is outlined in Figure 9.3–2. A suggested scheme to generate substructures is presented in Figure 9.3–3. This particular procedure is only intended to show how a substructure is prepared and shared in a general computational scheme.

The substructuring technique has been used to analyze a number of practical problems, and some of them listed below under the classification of repetitive structures, structures with singularities, structures with multiple solution steps, and large structures, assembled structures and multilevel systems.

Repetitive structures: Laminated plates, electrification towers, bonnet, stiffened sandwiches, 3–D sandwich fabric, and train wheel.

Structures with singularities: Variable thickness structures, joints, structures with holes, impact problems, notched beams.

Structures with multiple solution steps: Crushing of tubes, train wheel, local buckling.

Large structures, assembled structures, and multilevel systems: Mountain bike, foundations for posts, bus structure, trailer, truck frame, structural constructions.

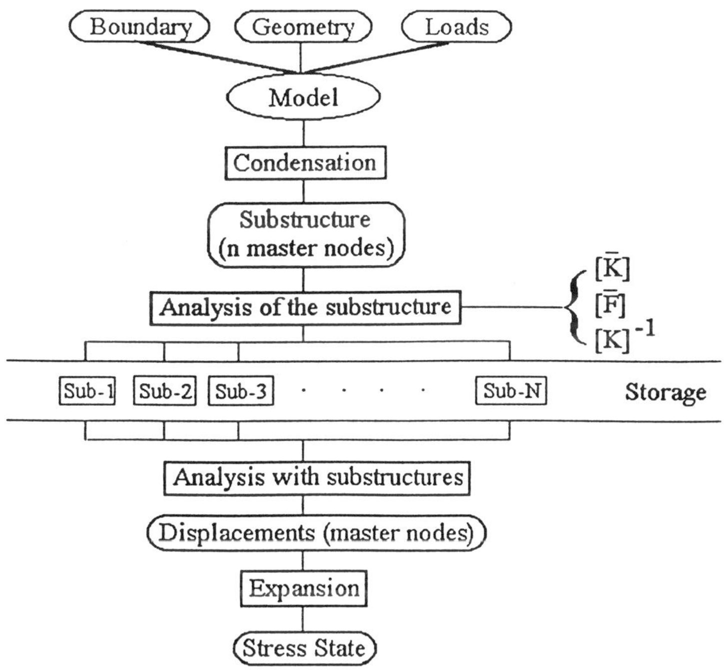

Figure 9.3–3: Generation, storage and managing of substructures.

9.4 THE DESIGN OF A BUS STRUCTURE

9.4–1 Introduction

One of the principal problems during the life of a vehicle made of metallic materials is the corrosion of some of its parts due to the environmental exposure, and therefore a continuous maintenance is necessary. Purely aesthetic damage and even weakening of the structure are encountered. Corrosion, which alters the mechanical characteristics of the materials of the vehicle and reduces their resistive capacity, makes it necessary to establish periodic repairs and replacement of parts, which in turn increase the total cost during the service life of the structure. Much work has been done in order to obtain vehicles with optimum durability characteristics (aesthetics, functional, etc.) over a period of time.

In most cases an increase in safety and low maintenance cost are recommended. Composite materials offer a wide spectrum of possible solutions since corrosion is not considered as a problem with composite materials. Parts of structures made of metals in machines, such as bumpers, hub caps, and body work components, which are subjected to an environmental exposure have been replaced by pieces made of nonmetallic materials. Other outstanding advantages of composite materials in the transportation industry are the weight savings obtained by the substitution of heavy parts by light pieces made of composite materials, which results in low fuel consumption, higher velocities, and increased payloads (passengers or goods). A significant parameter when designing with composite materials is their specific weight; many uses of composite materials in the automotive industry are due to this property.

9.4–2 Analytical Design of a Bus Structure

The analysis of the body structure of a bus made of composite materials is presented here. Several modifications to the structure of the current body of a bus have been made in order to guarantee the stability of the structure and safety of passengers in the case of a bus turn over. Usually reinforced metal structures make the structure strong but heavy, whereas composite material can be used to obtain a weight reduction. Continuous composite structures, as opposed to structures formed by discrete parts, are more suitable in the design of a modular structure like the bus structure. The structural continuity simplifies the use of rectangular frames, which represent a repetitive structure. The structure obtained this way reduces its weight and the height of its center of gravity, increasing the stability of the vehicle. Another important consequence of the weight reduction is the saving in fuel consumption. Figure 9.4–1 shows the modular character of the bus structure.

The material used in the design of the bus structure depends on the part being considered. Basically there are two types of components; the frame structure and the sandwich panels that are connected to the frame. Hybrid glass/carbon laminates were used in the continuous frame structure, and carbon was used in parts where stress concentration appears and in areas where high stiffness is required. Reinforced laminated sandwich panels are used for the floor structure of the bus, and nonreinforced laminated sandwich panels are used in the lateral portion of the body and in the roof of the body. Pultruded beams are used to join all the modules.

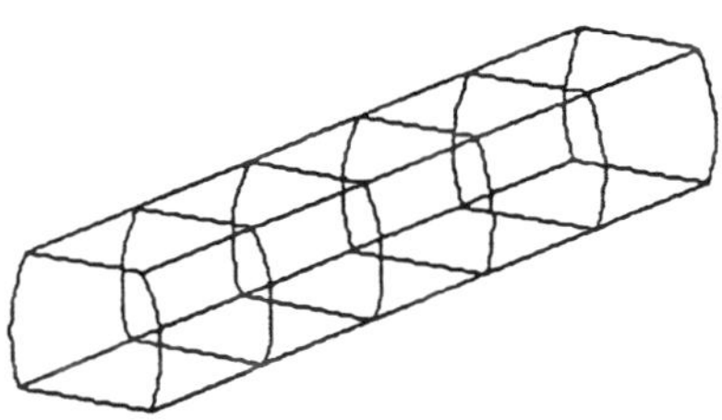

Figure 9.4–1: Modular structure of the bus.

The design of the body made of composite materials was realized by taking into account a large number of factors that make the process more complicated due to the increase in the number of variables. The bus structure made of composite materials must satisfy the same requirements as those of a bus made of metals. The composite structure must satisfy some additional requirements.

The stacking sequence, lamina thickness, and type of joint (adhesive or bolted) increase the number of variables of the problem greatly. However, there exists a wide range of possibilities to choose from appropriate material, manufacturing process, and resin and fiber systems.

A large structure like this can be successfully analyzed by subdividing it into parts that are representative. Figure 9.4–2 shows different substructures forming the whole bus structure. In the analytical model, different types of substructures – from structural components of the structure to sublaminates in the panels – were modeled. Several parts have been considered as substructures due to the possibility of optimizing them during the analysis; it is easier to change the type of substructure than to re–model the entire structure. Other parts of the structure were modeled as substructures because of their complicated geometry and unknown structural response. Different parts of the model were modeled by different persons, and, therefore, computing and modeling time savings were obtained.

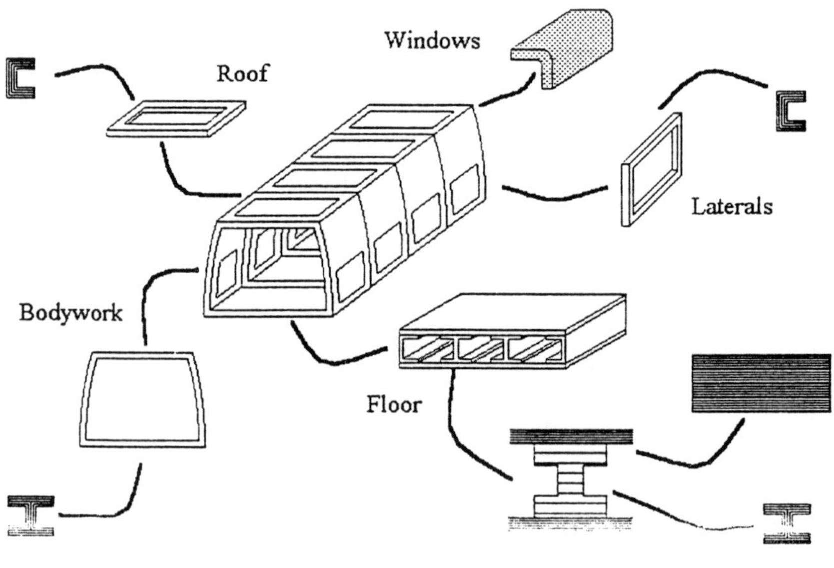

Figure 9.4–2: Bus substructure scheme.

9.4–3 Finite Element Analysis

The finite element analysis was performed by means of the substructuring technique. The analytical design process was supported by powerful workstations, and models were run in vectored and multi–tasked mainframe supercomputers, *e.g.*, CONVEX and HP. The ABAQUS finite element code and the IDEAS pre– and post–processor software were used to obtain stress distributions, deflections, and data for comparative purposes. A variety of finite elements was used in order to obtain the best configuration for each substructure, bearing in mind that different finite elements should be compatible between substructures that are joined. Simple models of each substructure were developed prior to the final modeling in order to know its behavior and to control the CPU time used and the computer storage.

An initial analysis of the whole structure gives us an idea of the behavior of the model (see Figure 9.4–3) as well as critical areas of the design. Once initial analysis was finished, the next step was to subdivide the model into substructures, each one representing parts of the structure. The analysis procedure for the bus structure was developed taking into account the contribution of the different parts. Components of the main structure were analyzed separately and optimized according to the general requirements of the design. Then, each part of the structure was brought together to form the whole model.

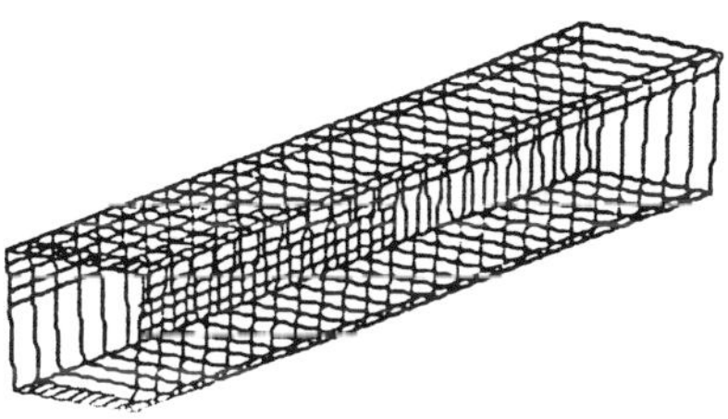

Figure 9.4–3: Initial model of the bus structure.

Some of the substructures used to represent the whole model are shown in Figures 9.4–4 through 9.4–6. The substructures shown in Figures 9.4–4 through 9.4–5 represent some examples of the substructures in the library of substructures created during the modeling process. As can be seen from Figure 9.4–4, mesh refinement was made in some parts of these substructures because of their expected behavior. Each substructure was modeled with different types of finite elements, depending on its own requirements and the desired degree of mesh refinement. Important considerations of the modeling were the node locations and element connectivity between different substructures. Therefore, a careful selection of nodes that joined the substructure to others was used in order to avoid discontinuities. A continuous communication between the persons modeling different substructures was established, and information about the position of boundary nodes was obtained at each stage of the modeling process.

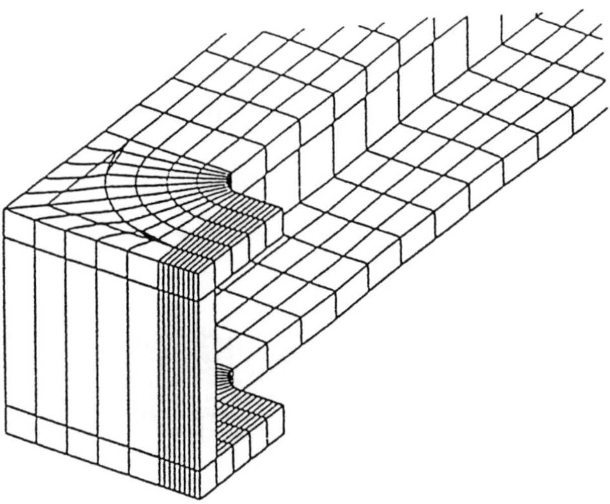

Figure 9.4–4: A part of frame substructure.

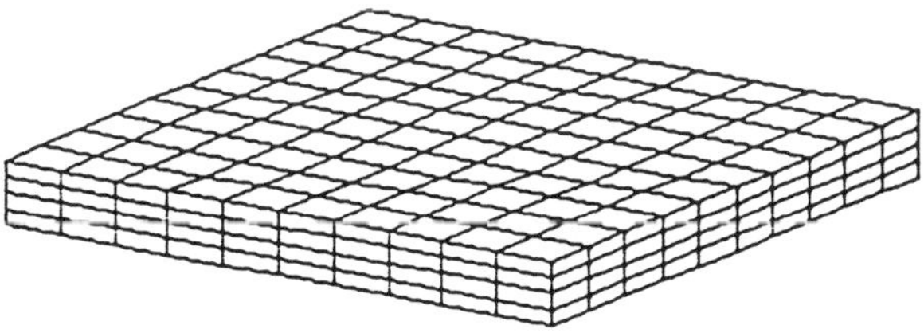

Figure 9.4–5: Laminated sandwich panel substructure.

Once the whole model was constructed, an exhaustive analysis was carried out and the first results were obtained. These results were returned to the different substructures, and suitable modifications were made on each one. An iterative process of optimization was established until the principal objective was realized. Once the iterative process ended, the results provided by the analysis were compared with real and experimental data, and an optimum design of the bus structure was obtained.

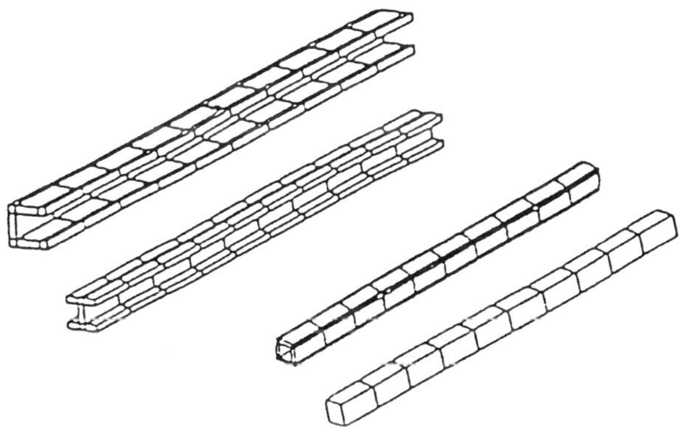

Figure 9.4–6: Stiffeners of the floor substructure.

9.4–4 Optimization

When a structure is composed of several different substructures, it is difficult to select the material of all substructures and parts and to fix dimensions in a first approximation. Therefore, an optimization of some of them is recommended in order to get a competitive design. Several parts of the structure were optimized depending on the objectives for which they were selected, while other parts of the model were fixed because of the design. Optimization was carried out to determine the appropriate laminate in both continuous frames and sandwich skins. The stiffeners in the sandwich structures have been optimized as well. Referring to the entire model, the number of modules as well as their location in the main structure have also been analyzed. Some examples of the results obtained by the procedure mentioned above are presented in Figures 9.4–7 through 9.4–10.

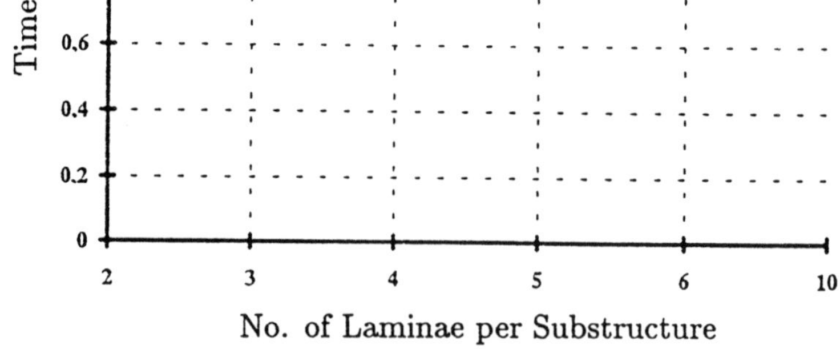

Figure 9.4–7: Optimization of a laminate.

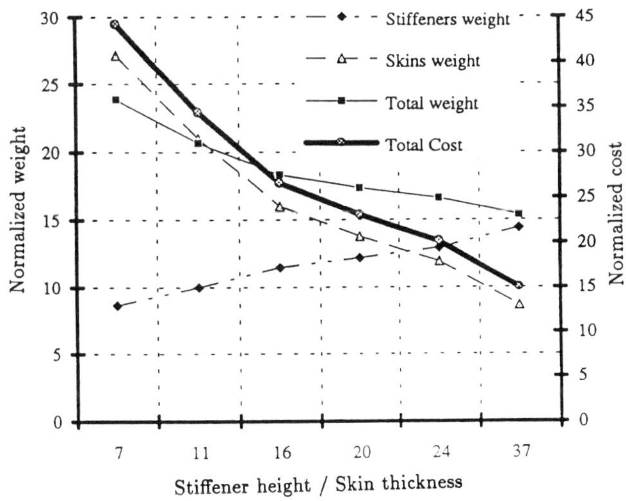

Figure 9.4–8: Optimization of a sandwich structure.

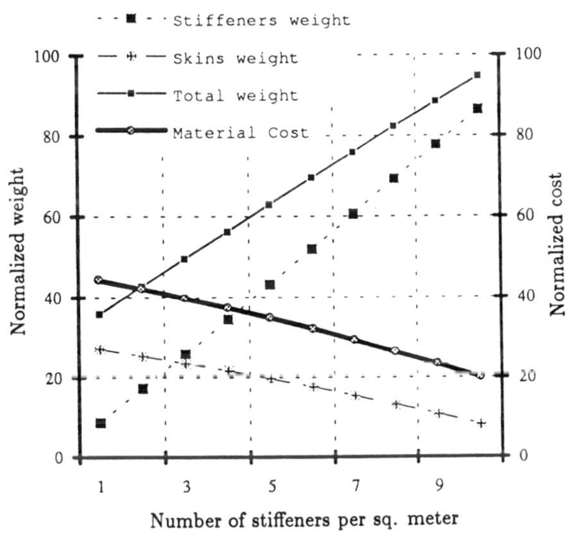

Figure 9.4–9: Optimization of number of stiffners.

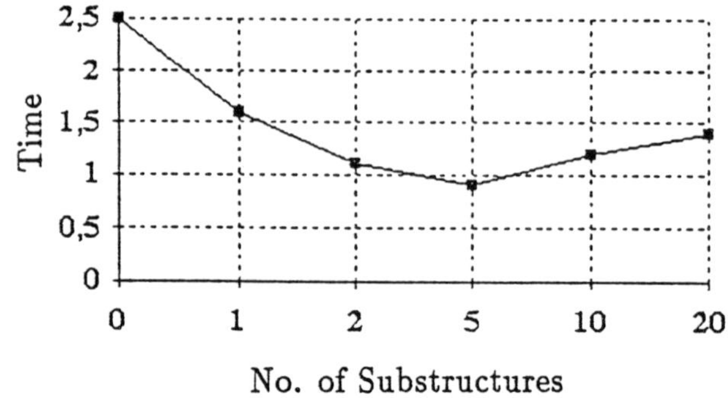

Figure 9.4–10: Optimization of the number of substructures in each module.

9.4–5 Conclusions

Finite element computer codes with the substructuring technique have been used to analyze structures made of composite materials. Due to the great number of variables involved in the problem, the substructuring technique was used. An important advantage of this technique is that it allows for subdivision of work as follows:

- Isolation of critical zones (mesh refinement on detailed studies).
- Separation among design variables and constants.
- Separation among stresses and displacements calculations.
- Separation among linear and nonlinear analysis.
- Selective changes can be made in specific substructures without a global remeshing.
- Efficient method to store substructures and to create an easy use of libraries.
- Branch system with several substructuring levels can be established.

- A work team can be created.

Savings in computational time and reduction of modeling effort can be obtained with this technique. The order of the system matrices is reduced, and the stiffness matrix of only the first element needs to be derived, which reduces the required computer effort. Large detailed structures can be analyzed in a very short period of time, and optimization process and sensibility analysis can be carried out. An effective method to analyze large structures (by means of a small model) can be achieved by making a library of substructures, and it is specially recommended in repetitive models, thus avoiding large computer storage and CPU times.

REFERENCES FOR ADDITIONAL READING

1. Bathe, K.-L., *Finite Element Procedures in Engineering Analysis,* Prentice-Hall, Englewood Cliffs, NJ, 1982.

2. Cook, R. D., Malkus, D. S., and Plesha, M. E., *Concepts and Applications of Finite Element Analysis,* Third Edition, John Wiley & Sons, New York, 1989.

3. Crisfield, M. A., *Non-Linear Finite Element Analysis of Solids and Structures,* John Wiley & Sons, Chichester, U.K., 1991.

4. Reddy, J. N., *An Introduction to the Finite Element Method,* Second Edition, McGraw-Hill, New York, 1993.

5. Zienkiewicz, O. C. and Taylor, R. L., *The Finite Element Method,* McGraw-Hill, New York, 1989.

6. ABAQUS, version 4.9, Hibbitt, Karlsson and Sorensen, Inc.

7. Mondkar D.P. and Powell G.H., "Large Capacity Equation Solver for Structural Analysis," *Computers and Structures,* **4**, 699–728, 1974.

8. Irons, B. and Ahmad, S., *Techniques of Finite Elements,* Ellis Horwood Ltd., Chichester, England, 1980

9. Kim J. Y. and Hong C. S., "Three-Dimensional Finite Element Analysis of Interlaminar Stresses in Thick Composite Materials," *Computers & Structures,* **40**(6), 1395–1404, 1991.

10. Liu Xia-shi, "Finite Element Analysis for Large Composite Structure," *Proceedings of ICCS IV,* Paisley, Scotland, 1987.

11. Larrodé, D. "Análisis de grandes estructuras realizadas en materiales compuestos por medio de técnicas de elementos finitos y de substructuración," Ph.D., University of Zaragoza, Spain, 1994.

12. Larrode, E., Miravetef, A., and Alba, J. J., "Analysis of Composite Structures Using Substructure Techniques," *Proceedings ICCM/9*, Vol. 3, pp. 169–182, A. Miravete (ed.), Madrid, 1993.

13. Larrode, E. and Miravete, A., "A Finite Element Analysis of a Bus Structure using Substructure Techniques," *Internationa Journal of Composite Structures*, to appear.

14. Larrode, E., Miravete, A., and Ullod, J., "Análisis mediante elementos finitos utilizando la técnica de las subestructureas de una estructura de autobús," *Proceedings del I Congreso Iberoamericano de Ingeniería Mecánica,* Madrid, 1993, Vol. 4, pp. 73–82.

Chapter Ten

PRACTICAL ANALYSIS WITH *DAC*

10.1 INTRODUCTION

10.1–1 General Comments

DAC (Designing Advanced Composites) is a user–friendly software package developed for the linear analysis of standard composite–material structural elements by the Composite Materials Group at the University of Zaragoza, Spain, under the supervision of Professor Antonio Miravete. *DAC* programs run on an IBM PC and compatibles under an MS–Windows environment. The user can work with a mouse, display several "windows" simultaneously on the screen, and use all the facilities that MS–Windows offers. Solutions of various problems are based either on the finite element method or analytical solutions of the classical laminate theory, and test results are also included wherever appropriate. All *DAC* programs have been developed with a similar user interface, which is carefully designed to be used by a inexperienced user. The material database is shared by all modules that use external material files. Optimization and sensitivity studies are also included, and the results are presented in a chart form, showing how results change with some of the input parameters. Thus, the user can find, for instance, the required thickness plate or the optimum ply orientation for maximum/minimum deflection or strength requirements in a few seconds.

10.1–2 Why *DAC* ?

Most of the available computer programs on composite materials and structures are large computer codes and are run on minicomputers or workstations. These programs are expensive and the computers needed to run them are also very expensive because of the memory requirements. A typical user of these general–purpose programs spends a great deal of time defining the problem even for a simple structure such as a rectangular plate or a sandwich panel. Moreover, when applying the finite element method to laminated composite structures, problems arise in the calculation of interlaminar stresses or stresses in the free–edge regions of the laminates. Three–dimensional finite element models can be used to determine accurately the interlaminar and free–edge stress fields. However, the three–dimensional analyses require enormous storage and computational time. Therefore, three–dimensional analyses are seldom used in practice. In addition, three–dimensional analyses are not warranted in most preliminary design considerations. Most preliminary designs require a quick theoretical information concerning the displacements, stress states, buckling loads, and natural frequencies. A software package based on the finite element method (with a fixed mesh), analytical solutions, or experimental results that would be cheaper and quick to use is developed for use on a personal computer. These considerations lead to the development of *DAC* software for use on personal computers. They are intended to calculate the most commonly required information (deflections, stresses, buckling loads, etc.) for basic structural elements, such as plates, sandwich panels, beams and grids, in a few seconds with a very high accuracy and low computational cost.

DAC programs also contain a unique optimization and sensitivity study routine. In this way, the *DAC* software package can be used for two different aspects. On one hand, these programs are accurate tools that aid the design of composite–material structures; on the other hand, one can use the programs to investigate the basic behavior of composite materials when intuition cannot be used because of the complicated effect of several parameters on the load–carrying capability. Such information improves the previous knowledge as a complement to the theory presented in the books on composite materials.

10.2 *DAC* SOFTWARE

10.2–1 Software and Hardware Requirements

The *DAC* family is composed of nine modules, each having a specific computational purpose. The following configuration is recommended to run the *DAC* programs properly.

- Software:
  - Microsoft DOS 5.0
  - Microsoft Windows 3.1
- Hardware:
  - IBM i386DX/33 MHz with math coprocessor
  - 4 Mb RAM
  - Hard disk with at least 5 Mb of free space
  - 800 × 600 SVGA graphics card (1 Mb)

The *DAC* programs can be purchased by writing to the authors. The programs will be provided with new releases. On–line help and manuals are provided with every module and a hot–line service is available for the user to resolve any doubt or to report any suggestion.

10.2–2 Computational Time

The *DAC* programs are very fast in computing the solutions. The computational time depends on the type of problem solved. Problems requiring finite element analysis require more computational time than those based on analytical solution or test results, which are instantaneous. Other factors, such as the number of plies (the maximum number of plies allowed is 200), number of reinforcements, number of plies for the reinforcement, and the number of points calculated for the sensitivity study, also influence the total solution time.

The computational times taken by the *DAC* modules for some typical example problems are presented in Table 10.2–1. The program *DAC Reinforced Sandwiches,* which is the slowest module, and *DAC Plates/Sandwiches* with *DAC Beams/Grids* are based on the finite element method, and *DAC Non–Rectangular Plates* is based on analytical formulation. In each case, the structure is optimized for the independent variable θ, the angle of orientation, ranging from 0° to 90° with a step of 10°. Ten points were calculated. The computational times required by PCs with 386 and 486 chips for the three problems described below are presented in Table 10.2–1.

- A simply supported square sandwich panel of inplane dimensions 1 m by 1 m, reinforced with 4 beams along the x–direction and 4 beams along the y–direction is analyzed using *DAC Reinforced Sandwiches* module. The program allows up to 11 × 11 reinforcements.
- A simply supported square laminate of inplane dimensions 1 m by 1 m and stacking sequence $[(0/90)_{20}]_{sym}$ (80 plies) is analyzed with *DAC Plates/Sandwiches* module.
- An equilateral triangular plate of side 1 m with all edges clamped and having the lamination scheme $[(0/90)_{20}]_{sym}$ (80 plies) is analyzed with *DAC Non–Rectangular Plates* module.

Table 10.2–1: Computational times taken by *DAC* (in seconds)

DAC Program	Input Data	One Solution	Optimization
	386/33MHz.		
DAC Reinforced Sandwiches	<120	6	58
DAC Plates/Sandwiches	< 60	3	29
DAC Non–Rectangular Plates	< 30	< 1	1
	486/33MHz.		
DAC Reinforced Sandwiches	< 120	2	19
DAC Plates/Sandwiches	< 60	1	12
DAC Non–Rectangular Plates	< 30	< 1	< 1

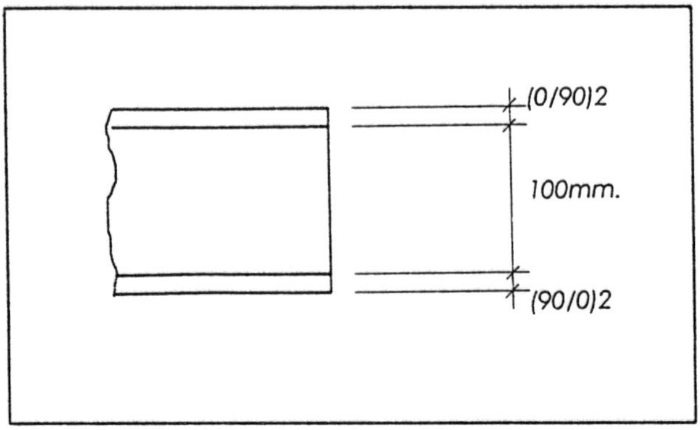

Figure 10.2–1: Sandwich structure.

10.2–3 Accuracy of the Results

Results obtained with *DAC* programs have been compared with exact results available in the literature (see, for example, Chapter 5) for isotropic and laminated structures, results obtained using other general–purpose finite element software, and laboratory test results. The results are found to be very accurate. The margin of error between results of a *DAC* program and the exact solution is less than 5%, working with conditions specified, such as sandwiches with the usual ratio of skin thickness to core thickness, specially orthotropic laminates, or laminates with enough repetitions of the sublaminate. A comparison of the results predicted by the *DAC* programs based on the finite element method is included in Tables 10.2–2 through 10.2–4 for the following example problems.

- A symmetric cross–ply laminate (0/90/0/90/0/90/0/90/0), simply supported and subjected to uniformly distributed transverse load, is analyzed for various values of the ratio of plate length to thickness a/h ($a = b$). The material used is HM graphite epoxy (see Table 10.2–2).

- A simply supported, square, angle–ply laminate (45/-45/45/-45/45/-45/45/-45/45) under uniformly distributed transverse load is analyzed for various values of the ratio of plate length to thickness a/h. The material used is HM graphite epoxy (see Table 10.2–3).

Table 10.2–2: Accuracy of deflections, $\bar{w} = w(E_2h^3/q_0a^4) \times 10^3$, at the center of symmetric cross–ply laminated plate under uniformly distributed traverse load.

a/h	DAC	Exact	Error (%)
10	6.04	5.85	3.2
100	4.55	4.49	1.3
1000	4.55	4.47	1.8

Table 10.2–3: Accuracy of deflections, $\bar{w} = w(E_2h^3/q_0a^4) \times 10^3$, and stress resultants at the center of the angle–ply laminated plate under uniformly distributed transverse load.

Quantity	a/h	DAC	Exact	Error (%)
Deflection				
	10	8.40×10^{-7}	8.13×10^{-7}	3.24
	100	5.17×10^{-3}	5.22×10^{-6}	-0.96
	1000	0.5176	0.5197	-0.39
Stress Resultant				
	10	0.8749	0.8783	0.13
	100	87.948	91.125	-3.61
	1000	87.948	90.871	-3.32

- Simply supported and clamped (on all edges) square laminates, for various stacking sequences, and subjected to uniform line load through the center of the plate, are analyzed. The material used is unidirectional glass fiber R 1200 (see Table 10.2–4).

- A simply supported, square, isotropic (steel) plate and subjected to uniformly distributed transverse load over a square patch (for different sizes of the patch) at the center, are analyzed. The material used is unidirectional glass fiber R 1200 (see Table 10.2–5).

Table 10.2–4: Accuracy of deflections, $\bar{w} = w(E_2h^3/q_0a^4) \times 10^3$, at the center of simply supported and clamped (on all four edges) laminated plates with various stacking sequences.

Laminate	*DAC*	Exact	Error (%)
Simply Supported Plate			
$0°$	5.01×10^{-3}	5.09×10^{-3}	1.70
$(0°/90°)_s$	3.04×10^{-3}	3.07×10^{-3}	1.07
$(0°/90°/45°/-45°)_s$	2.99×10^{-3}	3.07×10^{-3}	2.67
Clamped Plate			
$0°$	1.01×10^{-3}	1.02×10^{-3}	0.87
$(0°/90°)_s$	6.14×10^{-4}	6.15×10^{-4}	0.13
$(0°/90°/45°/-45°)_s$	6.04×10^{-4}	6.15×10^{-4}	1.70

Table 10.2–5: Accuracy of deflections, $\bar{w} = w(E_2h^3/q_0a^4) \times 10^3$, at the center of an isotropic plate subjected to uniformly distributed transverse load.

DAC	Exact	Error (%)
8.62	8.56	0.73
22.60	22.56	0.21
30.33	28.99	4.61
41.73	40.67	2.61
48.00	47.99	0.01

10.3 *DAC* PLATES/SANDWICH PROGRAM

This is a finite element analysis program that solves the deflection and strength ratio (Tsai-Wu criterion) of a rectangular symmetric plate or sandwich laminate, having free, simply supported or clamped boundary conditions on two opposing sides of the plate. Load can be a centrally located point load, a uniform pressure over the entire plate and/or any rectangular area. The sandwich core can be any material and thickness. Since a shear deformation theory is used, the shear modulus of the core can be less than perfectly rigid. Calculations are made for first–ply–failure (FPF) and last–ply–failure (LPF) theories. This is a very fast program and offers a unique optimizations routine and cost and weight estimation. *DAC* Plates/Sandwiches offers all the options described below.

Geometry:

- Rectangular plates with symmetric laminate.
- Rectangular sandwich panels. The upper skin is always symmetric to the lower one. Two different materials can be assigned, one for the skins and another for the core.

Boundary conditions:

- Two simply supported edges plate.
- Simply supported plate.
- Two clamped edges plate
- Simply clamped plate.
- Two opposite clamped edges and two opposite simply supported edges.

Load cases:

- Centrally located point load.
- Uniform pressure over the entire plate.

- Uniform pressure over a defined rectangle centered on the plate.
- Combination of the previous load cases.

Materials:

- Laminate materials for plates and skins of the sandwich.
- Core materials for the sandwich.

  Some materials files are included, but the user can enter the desired new materials. As a particular case, it is possible to enter isotropic materials.

Independent variables for an optimization or sensitivity study:

An optimization analysis or a sensitivity study shows the variation of the program outputs when one or two independent variables change. These variables can be chosen from the list below:

- Ply orientation angle for each ply group used to define the sublaminate of the plate or sandwich skin.
- Number of plies for each one of these ply groups.
- Number of times that the sublaminate is repeated.
- Angle rotation for the laminate or fabric.
- Thickness of the sandwich core.
- Dimensions of the plate: length and width.
- Numerical values of the loads and dimensions of the rectangular area load.

Outputs:

For a single calculation case, the available outputs are as follows:

- Maximum deflection for intact material properties (FPF).
- Maximum deflection for degraded material properties (LPF).

- Strength ratio for FPF.
- Strength ratio for LPF.
- Weight of the plate or sandwich.
- Cost.
- In–plane stress distribution diagrams and interlaminar shear stress distribution diagrams (when thick plate or sandwich) at the center of the plate, the midpoint of side X and the midpoint of side Y. For an optimization or sensitivity study, the available outputs are as follows:

- Charts showing how the maximum deflection FPF and LPF, strength ratio FPF and LPF, weigth, and cost change as the independent variables vary.
- Numerical results showing how the maximum deflection FPF and LPF, strength ratio FPF and LPF, weight, and cost change as the independent variables vary.

10.4 *DAC* REINFORCED SANDWICHES PROGRAM

This is a finite element analysis program that allows the calculation of deflections and the strength ratio (using the Tsai–Wu criterion; see Chapter 6) of a rectangular symmetric sandwich panel reinforced with composite material profiles. Profiles can be placed in some defined positions along the x and y coordinates, always in a symmetric way. Profile sections can be easily chosen among different kinds, and their material and laminate stacking can be defined. The load can be (a) a centrally located point load, (b) a uniformly distributed (pressure) transverse load over the entire sandwich, and/or (c) a uniformly distributed transverse load over any rectangular area. As in *DAC* plates/sandwiches, the sandwich core can be any material and thickness, and the shear modulus of the core can be less than perfectly rigid because a shear deformation theory is used. Calculations are made for FPF and LPF theories. Cost and weight estimation and optimization or sensitivity studies can also be carried out. *DAC* Reinforced Sandwiches offers all the options described below:

Geometry:

- Rectangular sandwich panels. The upper skin is always symmetric to the lower one. Two different materials can be assigned, one for the skins and another for the core.
- Rectangular plates with symmetric lamination schemes.

Boundary conditions:

- Two opposite edges simply supported and the other two free.
- All four simply supported edges.
- Two opposite edges clamped and the other two free.
- All four edges clamped.
- Two opposite edges clamped and the other two opposite edges simply supported.

Load cases:

- Point load on the center of the plate.
- Uniform pressure over the entire plate.
- Uniform pressure over a defined rectangle centered on the plate.
- Combination of the previous load cases materials.
- Laminate materials for plates, skins of the sandwich and reinforcements.
- Core materials for a sandwich.

  Some materials files are included but the user can enter the desired new materials. As a particular case, it is possible to enter isotropic materials.

Reinforcements:

The library of beam cross sections include the following:

- Rectangular section.
- Box section.
- I–section.
- C–section.

The beam section data consist of the following:

- Section height.
- Section width.
- Thickness and/or laminate definition.
- Material.

The user can also enter a general section defined by means of rectangles.

Reinforcements:

An optimization analysis or a sensitivity study shows the variation of the program outputs when one or two independent variables change. These variables can be chosen from the list below:

- Ply orientation angle for each ply group used to define the sublaminate of plate or sandwich skin.
- Number of plies for each one of these ply groups.
- Number of times that the sublaminate is repeated.
- Angle rotation for the laminate or fabric.
- Thickness of the sandwich core.
- Dimensions of the plate (length and width) and at the same time length of the reinforcements placed along these directions.
- Numerical values of the loads and dimensions of the rectangular area load.

Outputs:

For a single calculation case, the available outputs are as follows:

- Maximum deflection for intact material properties (FPF).
- Maximum deflection for degraded material properties (LPF).
- FPF strength ratio for the sandwich panel or plate.
- LPF strength ratio for the sandwich or plate.
- The lowest of the FPF strength ratios for the reinforcements placed along the $x-$direction.
- The lowest of the LPF strength ratios for the reinforcements placed along the $x-$direction.
- The lowest of the FPF strength ratios for the reinforcements placed along the $y-$direction.
- The lowest of the LPF strength ratios for the reinforcements placed along the $y-$direction.
- Global FPF strength ratio for the structure.
- Global LPF strength ratio for the structure.
- Weight of the structure.
- Cost of the structure.

For an optimization/sensitivity study, the outputs are available in a chart form as well as in numerical form for the following quantities:

- Maximum deflection for original material properties.
- Maximum deflection for degraded material properties.
- Global FPF strength ratio for the structure.
- Weight and cost change as the independent variables vary.

10.5 SAMPLE APPLICATIONS

10.5–1 Advertisement Sign Board

Analysis was performed to develop an optimal structure for the advertisement board shown in Figure 10.5–1. The board has the following dimensions: height (width), b = 2.5 m and length, a = 6 m. The wind velocity of 50 m/s is represented as a uniformly distributed transverse load of 1,000 N/mm^2 on the structure.

Figure 10.5–1: An advertisement sign board.

The objective is to obtain a structure that has a stiffness such that the board will not deflect more than 10 mm under the load. As a design criterion, the following engineering option has been used: the maximum displacement should be less than the width of the structure divided by 250.

First a board without a core was considered. For this case, a polyester composite reinforced by a fabric of glass fibers was assumed. To determine the optimal layup for the manufacturing process, the properties of a unidirectional composite of polyester reinforced by glass fibers were used. For several different percentages of fibers in 0° and 90° the optimal layup of the fabric has been determined as a function of the length of the advertisement board, having a fixed width of 2.5 m (see Figure 10.5–2). The vertical line in the figure indicates the optimal layup for the dimensions used for the case. It can be seen that the optimal layup in this case is to orient all the fibers in 90°. This is the layup with the smallest displacement. Therefore, one must choose a layup with the maximum number of fibers with 90° orientation.

For further calculations, the following properties of a unidirectional composite were used:

Polyester reinforced by glass fibers:

Fiber volume (%)= 52

Ply thickness (mm)=0.5

Density (Kg/m^3)= 1926

E_1 (GPa)=39.29

E_2 (GPa)=10.28

G_{12} (GPa)=4.14

ν_{12}=0.26

Foam of polyurethane:

Density (Kg/m^3)=40

E_1 (GPa)= 8

E_2 (GPa)= 8

G_{12} (GPa)=3.2

ν_{12}=0.7

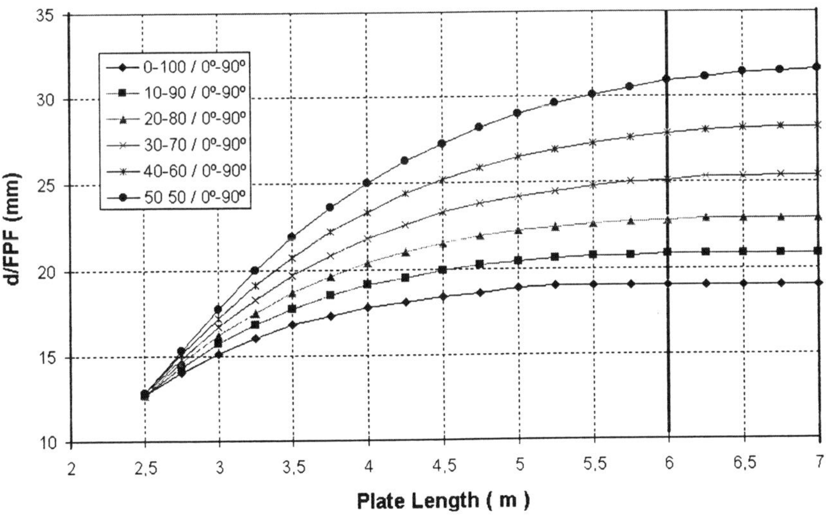

Figure 10.5–2: The influence of the plate length on the orientation of the fibers (without a core).

Figure 10.5–3 contains a plot of the maximal displacement as a function of the thickness for the advertisement board with dimensions of 2.5 × 6 m under the uniformly distributed transverse load.

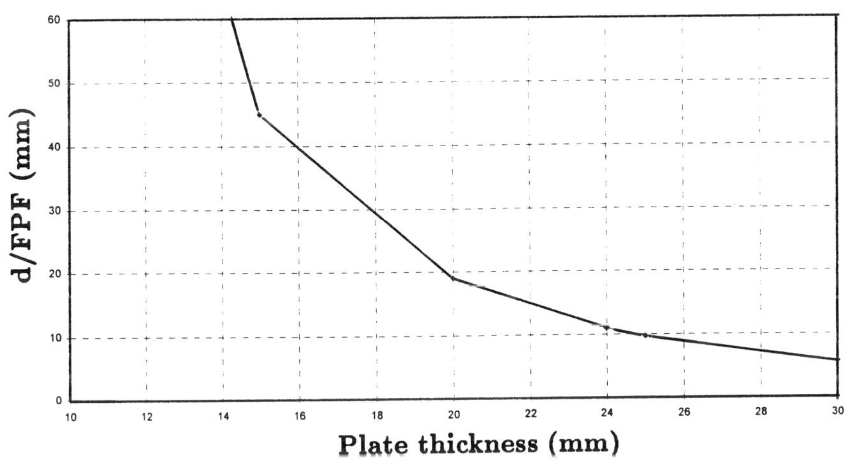

Figure 10.5–3: Maximum displacement (without a core).

With the above calculations it is clear that it is more useful to use a core in order to reduce the total number of plies needed for this structure. First, a core with a thickness of 20 mm was considered. For this case also the best solution is obtained for the case in which all fibers are oriented at 90° (see Figure 10.5–4). From Figure 10.5–5 it follows that the total laminate thickness should be at least 12 mm.

Next, a core with a thickness of 40 mm was considered. As one can see from Figure 10.5–6, the best solution is obtained for the case in which all fibers are oriented at 90°. Figure 10.5–7 shows that, with a core of 40 mm, plies with a total thickness of at least 3.5 mm are needed.

Given the ideal thickness of the plies, one likes to determine the minimum thickness of the core. Figure 10.5–8 shows that for a total thickness of 3 mm the thickness of the core must be at least 42 mm. Therefore, it can be concluded that a plate with a ply thickness of 3 mm (top and bottom) and a core thickness of 42 mm will satisfy the design requirements. Figures 10.5–9 through 10.5–14 show the stress distributions at the points indicated in Figure 10.5–15 for a plate with a total ply thickness of 3 mm and a core thickness of 42 mm.

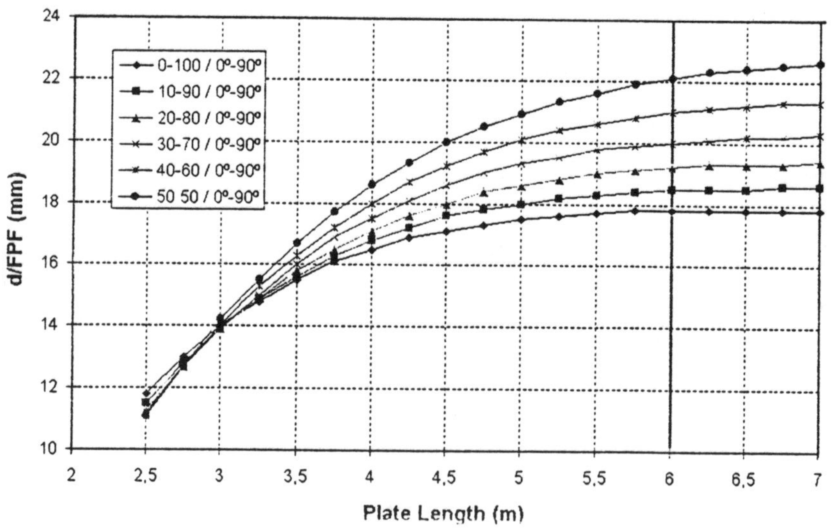

Figure 10.5–4: The influence of the plate length on the orientation of the fibers (with a core of 20 mm).

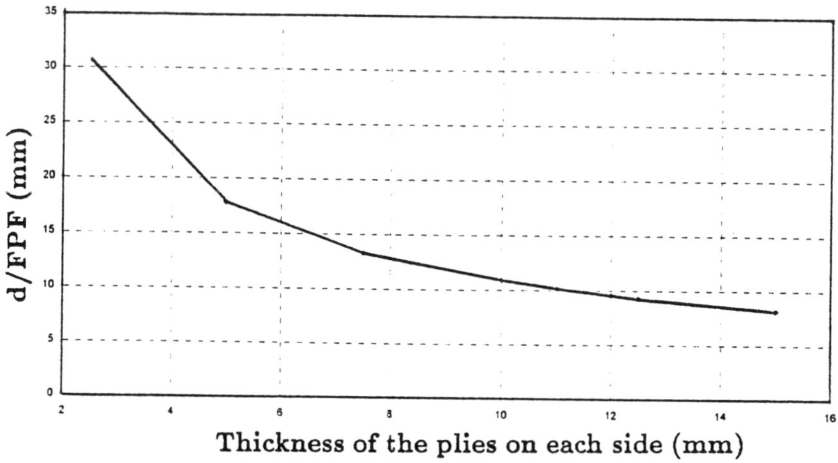

Figure 10.5–5: Maximum displacement (with a core of 20 mm).

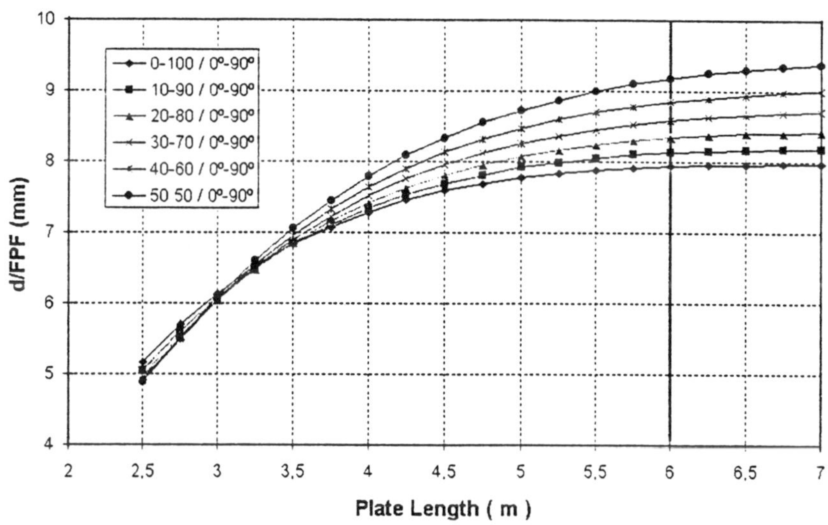

Figure 10.5–6: The influence of the plate length on the orientation of the fibers (with a core of 40 mm).

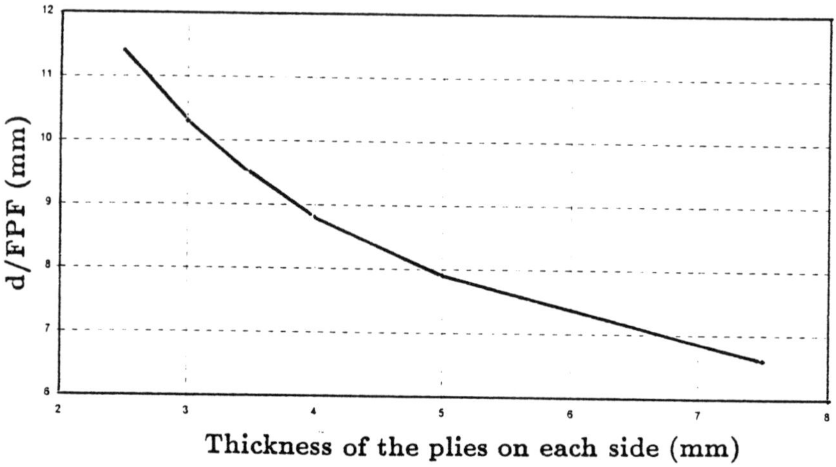

Figure 10.5–7: Maximum displacement (with a core of 40 mm).

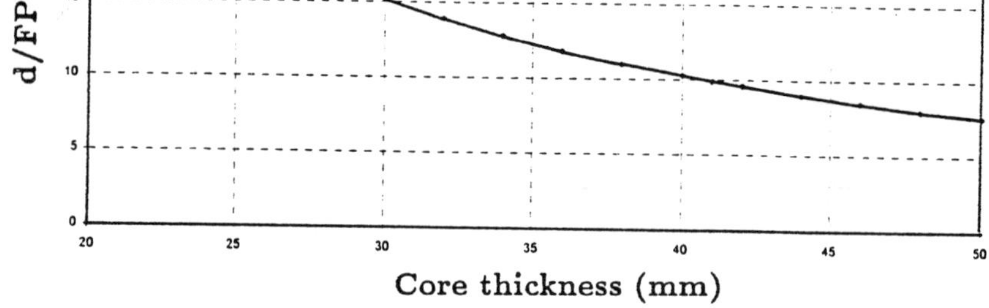

Figure 10.5–8: Maximum displacement as a function of core thickness for a plate with a total thickness of 3 mm.

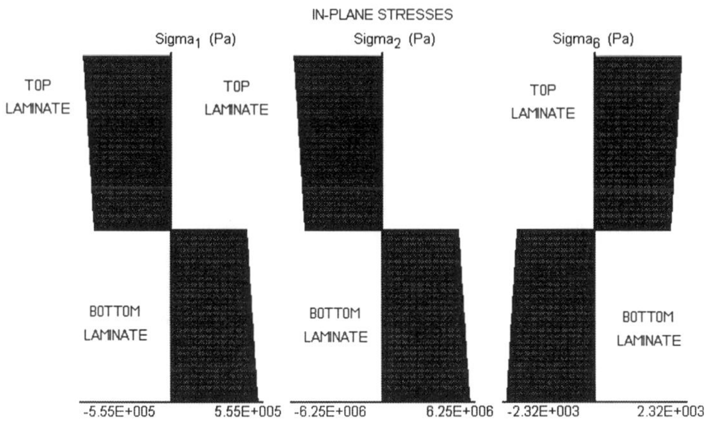

Figure 10.5–9: Inplane stresses at the plate center.

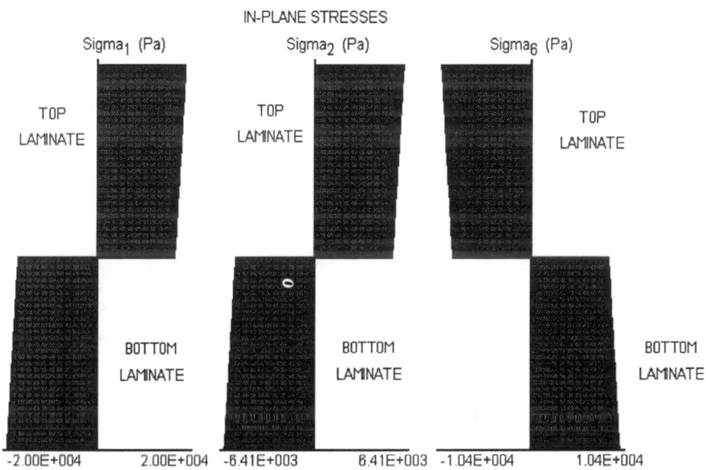

Figure 10.5–10: Inplane stresses at the midpoint of the side $x = a/2$.

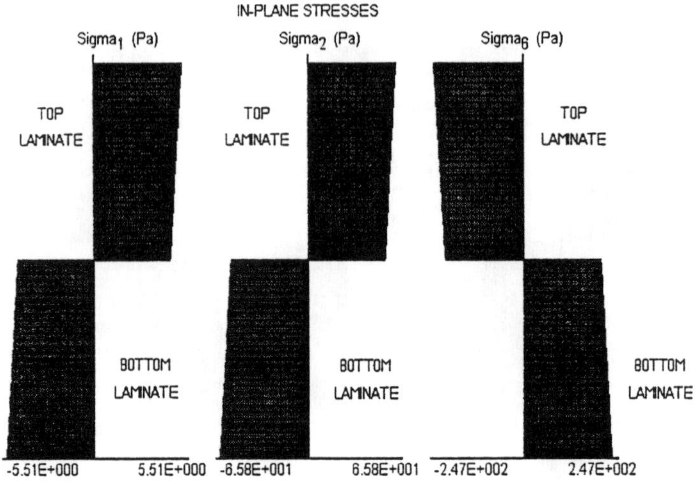

Figure 10.5–11: Inplane stresses at the midpoint of the side ($y = b/2$).

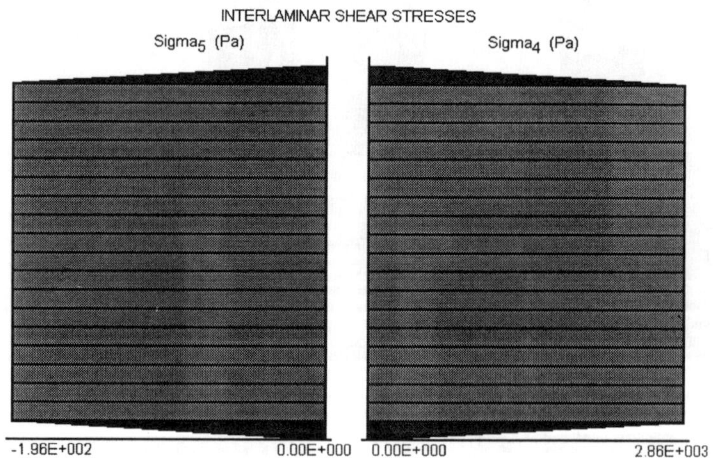

Figure 10.5–12: Interlaminar shear stresses at the plate center.

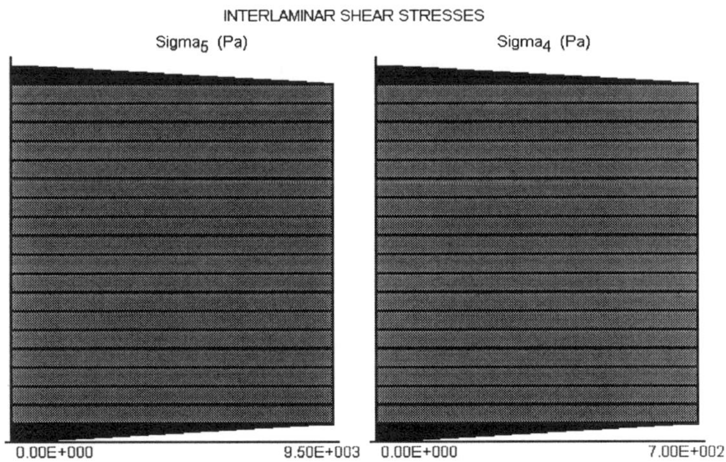

Figure 10.5–13: Interlaminar shear stresses at the midpoint of the side $x = a/2$.

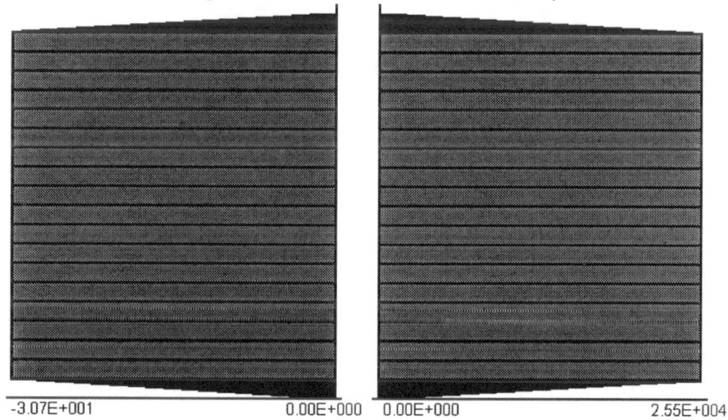

Figure 10.5–14: Interlaminar shear stresses at the midpoint of the side $y = b/2$.

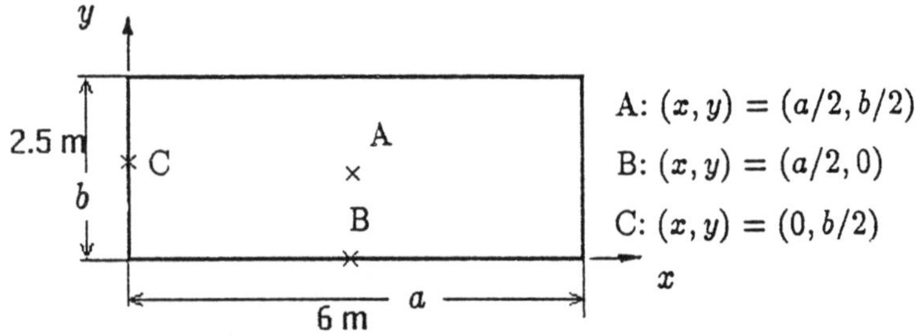

Figure 10.5–15: Location of the points where stresses were computed.

Figure 10.5–16: Trailer truck.

10.5–2 Trailer Floor

In this example, the floor of a trailer truck is analyzed (see Figure 10.5–16). The truck floor is assumed to be constructed with the same materials as used for the advertisement (or sign) board. Because of the magnitude of the loads that this structure has to support, a plate made of a sandwich reinforced by a series of beams in the transverse direction is considered. The dimensions of the floor of the truck are: length, $a = 12$ m and width, $b = 2.5$ m. The floor structure should be capable of supporting a uniform load of 10 kN/m^2 with a maximum transverse deflection of 10 mm.

The object of this study is to optimize the thickness of the plies, and determine the geometry of the reinforcements and the number of reinforcements such that the structure has the lowest weight possible.

Figures 10.5–17 through 10.5–24 show plots of the weight versus the distance between the reinforcements for several thicknesses of the reinforcements. One can see from these figures that when the distance between reinforcements increases, the thickness of the plies increases as well. The weight of the structure decreases in the first part and increases afterwards. At the first glance, the figures show that, for reinforcements with small thicknesses, the lightest structure can be found for small distances between the reinforcements. For reinforcements with greater thicknesses the lightest structure can be found for greater distances between the reinforcements, and the weight of the structure is greater than in the previous case. However, it is clear that there are many solutions for this problem. When the optimal solution for each reinforcement thickness is plotted (see Figure 10.5–25), it becomes clear that the minimum weight occurs for a distance between reinforcements equal to 0.2 m and a total thickness of 1 mm. This distance between two reinforcements is rather small, implying that the number of reinforcements is rather large. For a thickness of 2 or 3 mm the number of reinforcements decreases considerably (or the distance between reinforcements increases), and the weight remains almost the same or increases slightly. Therefore, the optimal solution will correspond to a thickness between 2 and 3 mm.

10.6 CLOSURE

The two examples presented here illustrate the capabilities of *DAC* programs. The user–friendliness coupled with the interactive and fast computational capabilities of the programs make them a valuable tool for obtaining preliminary design data and gaining a fundamental understanding of how various parameters (*e.g.*, materials, stacking sequence, geometry, boundary conditions, and loads) influence the results. Persons interested in the programs should contact the authors (within Europe, write to the second author; and in the Americas, contact the first author).

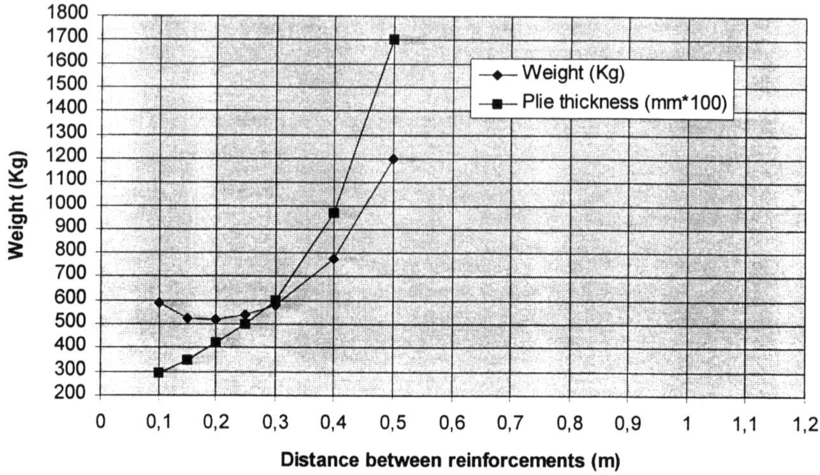

Figure 10.5–17: Plot of the weight versus the distance between reinforcements (reinforcement thickness = 1 mm).

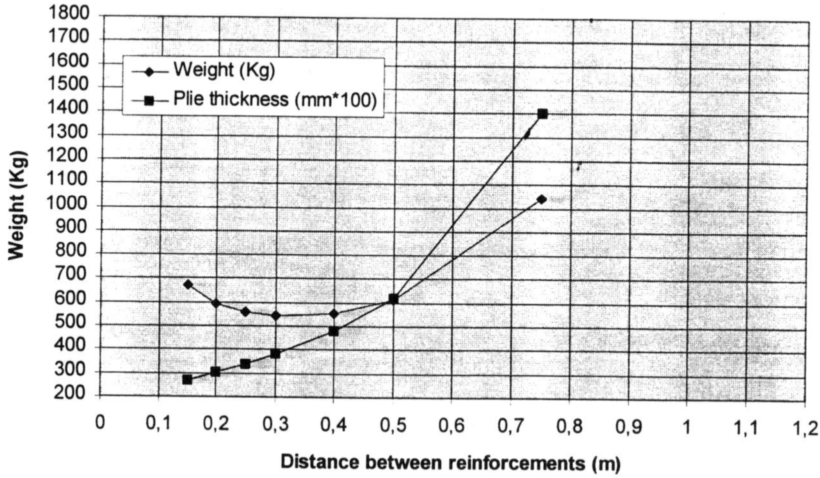

Figure 10.5–18: Plot of the weight versus the distance between reinforcements (reinforcement thickness = 2 mm).

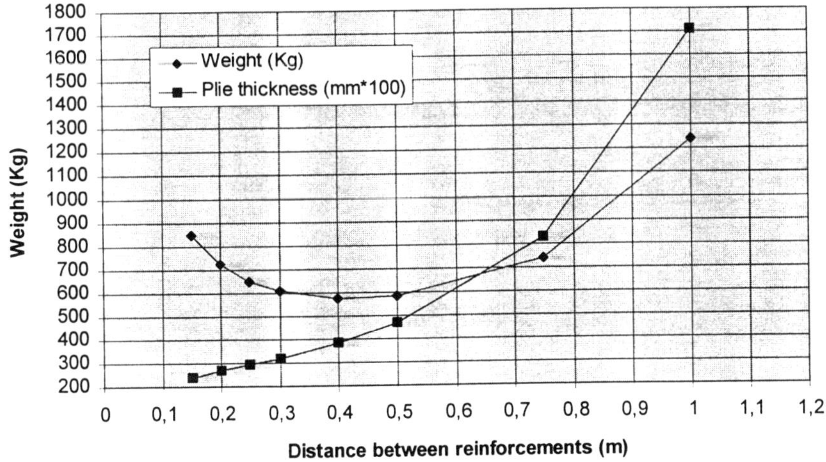

Figure 10.5–19: Plot of the weight versus the distance between reinforcements (reinforcement thickness = 3 mm).

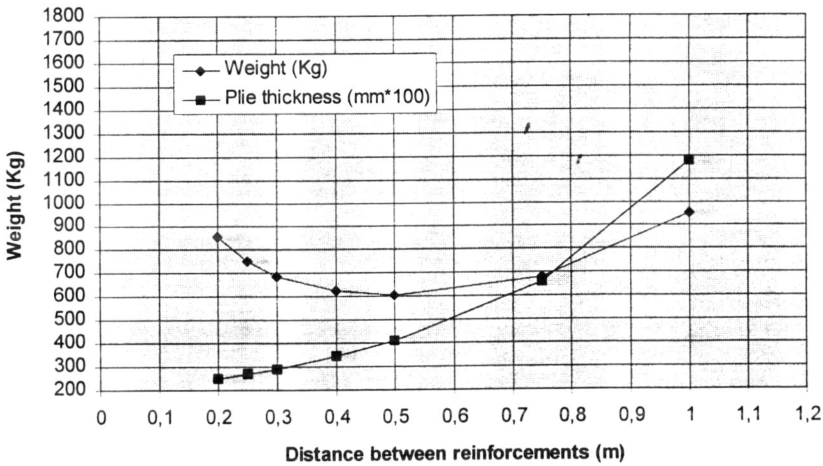

Figure 10.5–20: Plot of the weight versus the distance between reinforcements (reinforcement thickness = 4 mm).

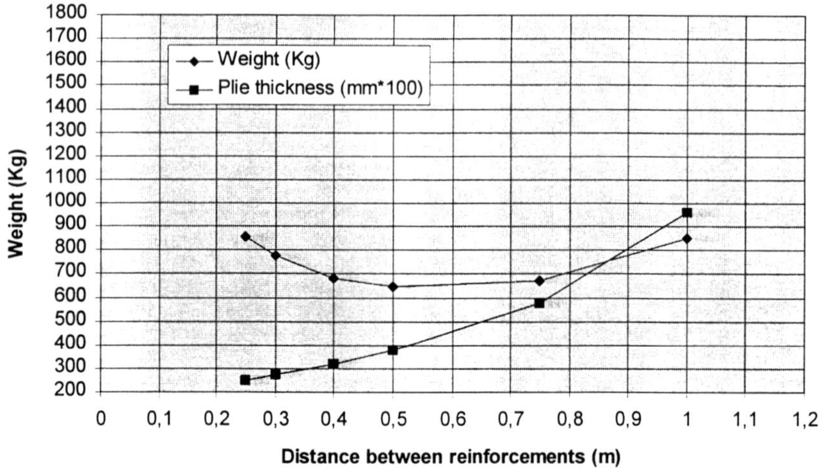

Figure 10.5–21: Plot of the weight versus the distance between reinforcements (reinforcement thickness = 5 mm).

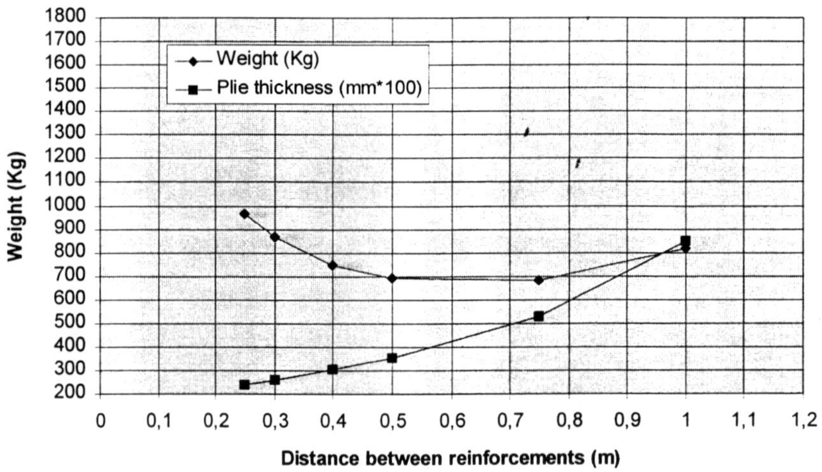

Figure 10.5–22: Plot of the weight versus the distance between reinforcements (reinforcement thickness = 6 mm).

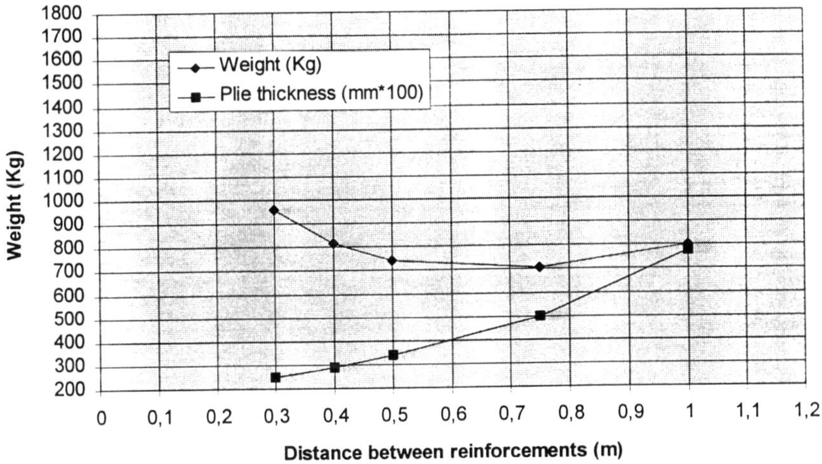

Figure 10.5–23: Plot of the weight versus the distance between reinforcements (reinforcement thickness = 7 mm).

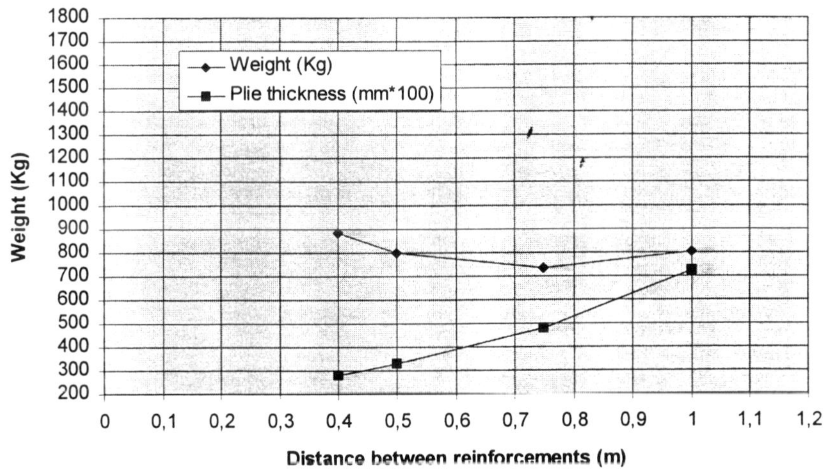

Figure 10.5–24: Plot of the weight versus the distance between reinforcements (reinforcement thickness = 8 mm).

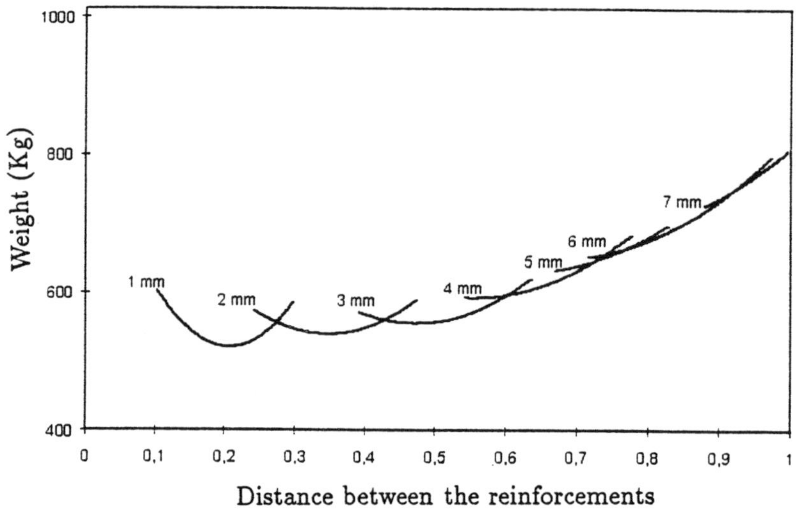

Figure 10.5–25: Plot of the weight versus the distance between reinforcements for various thickness of the reinforcement.

REFERENCES FOR ADDITIONAL READING

1. ABAQUS, Version 4.9, Hibbitt, Karlsson and Sorensen, Inc., Pawtucket, RI.

2. Jones, R. M., *Mechanics of Composite Materials,* Scripta Book Co., Washington, D.C., 1975.

3. Reddy, J. N., *Energy and Variational Methods in Applied Mechanics*, John Wiley & Sons, New York, 1984.

4. Reddy, J. N., *Mechanics of Composite Materials: Theory and Analysis,* John Wiley & Sons, New York, 1996.

5. Tsai S. W., *Theory of Composite Design,* Think Composites, Dayton, OH, 1992.

6. *DAC User's Manual,* Department of Mechanical Engineering, University of Zaragoza (C/ Maria de Luna, 3, 50015, Zaragoza), Spain, 1991.

7. Larrode, E., Casamayor, M. J., Miravete, A., and Ullod, J., "Design and Optimisation of a Full Composite Isotherm Structure," *International Journal of Composite Structures*, London, UK.

8. Larrode, E., Miravete, A., and Casamayor, M. J., "Optimización de un semirremolque isotermo," *Proceedings del I Congreso Iberoamericano de Ingeniería Mecánica*, Madrid, 1993, Vol 4, pp. 49–63.

9. Larrode, E., Miravete, A., Castejón, and L., Diseño, "Optimización y Cálculo de un elemento rodante en materiales compuestos para su aplicación al tren de alta velocidad español," *Proceedings del I Congreso Iberoamericano de Ingeniería Mecánica*, Madrid, 1993, Vol. 4, pp. 65–74.

SUBJECT INDEX